OXFORD MEDICAL PUBLICATIONS

Disability prevention: the global challenge

Disability prevention:
the global challenge

Edited by
SIR JOHN WILSON
President, International Agency for the Prevention of Blindness;
Director, Royal Commonwealth Society for the Blind

The contributors to this book wish to acknowledge the indispensable contribution made to its compilation and editing by Mr Kenneth Thompson, who was rapporteur at the Leeds Castle International Seminar.

No other single factor can contribute as much to diminishing the impact of disability as first-level prevention. Attempts to cure, restore or rehabilitate rarely give totally satisfactory results . . . First-level prevention should be the overriding priority for all national health authorities and organizations; and for WHO.

Expert Committee on Disability Prevention and Rehabilitation
WHO February 1981

Published on behalf of the Leeds Castle Foundation by
OXFORD UNIVERSITY PRESS
OXFORD NEW YORK TORONTO
1983

Oxford University Press, Walton Street, Oxford OX2 6DP

London Glasgow New York Toronto
Delhi Bombay Calcutta Madras Karachi
Kuala Lumpur Singapore Hong Kong Tokyo
Nairobi Dar es Salaam Cape Town
Melbourne Auckland

and associates in
Beirut Berlin Ibadan Mexico City Nicosia

Oxford is a trade mark of Oxford University Press

British Library Cataloguing in Publication Data
Disability prevention. — (Oxford medical publications)
1. Handicapped — Congresses
I. Wilson Sir John, 1919 –
362.4   HV1551
ISBN 0-19-261375-8

Library of Congress Cataloging in Publication Data
Main entry under title:
Disability prevention.
(Oxford medical publications)
Bibliography: p.
Includes index.
1. Medicine, Preventive — Congresses.   2. Preventive
health services — Congresses.   3. Handicapped — Congresses.
I. Wilson, John, Sir, 1919 –   .   II. Series.
[DNLM: 1. Preventive medicine — Congresses.   2. Handicap-
ped — Congresses.   WA 108 DG11]
RA422.D57 1983   362.4'047   83-2189
ISBN 0-19-261375-8

Set by Hope Services, Abingdon
Printed and bound in Great Britain by
William Clowes (Beccles) Limited, Beccles and London

Foreword

by the Rt. Hon. The Lord Home of the Hirsel K.T., D.L.

When I was Commonwealth Secretary in the late-1950s, I saw at first hand the ravages of tuberculosis and smallpox, blindness, polio, and leprosy from which millions in Africa and Asia suffered and died. It happened that, earlier in my life, I was totally immobilized for more than two years by tuberculosis of the spine. So I know something about disability and was pleased to accept an invitation to be chairman at an international seminar on the prevention of disablement convened at Leeds Castle in November 1981, towards the end of the International Year of Disabled Persons.

The International Year succeeded in changing public attitudes to disability and in removing barriers which isolate handicapped people, often unintentionally, from the life and activities of their community. Less was heard about another major purpose, disability prevention. The Trustees of the Leeds Castle Foundation and the British Ministry of Health and Social Security jointly carried out a suggestion from Sir John Wilson that a team of individuals should be invited to discuss this aspect on a world-wide, multidisciplinary basis and prepare a plan of action. Accordingly, twenty-five scientists, clinicians, Ministers of Health, and United Nations administrators, known to be leading men and women in their field internationally, spent five days together in the serene environment and seclusion of Leeds Castle in Kent. Each contributed of his knowledge and experience in combatting major causes of disablement. Together, they discussed the prospects and the means for initiating or reinforcing action to rid the world of preventable disabilities. Their recommendations for action, summarized in The Leeds Castle Declaration on the Prevention of Disablement (p. xi) were presented in December to the General Assembly of the United Nations and have been adopted by leaders of international agencies with a view to concerted action.

This book contains much of the material presented at the seminar, some of it rewritten by the participants in the light of the discussion at Leeds. It shows in more detail than the Declaration could do that most of the world's disablement is preventable and that much of it is curable. It shows also that the cost of effective action is not as great as might be supposed, provided that it is a concerted effort and not a series of independent campaigns. The opportunity now before us to reduce drastically the human waste and economic loss caused by avoidable disabilities is a challenge to us all.

Leeds Castle and Lady Baillie's Foundation

by the Rt. Hon. Lord Geoffrey-Lloyd, Chairman,
The Leeds Castle Foundation

I like to think of Leed, the Saxon Thane, riding along the gentle valley of the River Len some eleven centuries ago and suddenly coming upon a large lake formed by the little river in a natural hollow. He must have been a clever and powerful man as Chief Counsellor of Ethelbert IV, King of Kent, brother of Alfred the Great; but, with Danish invaders around, he must always have been thinking of the safety of his family. He saw that there were, surprisingly, two rocky islands in the lake and gradually, one supposes, he realized that he could make them into a safe refuge from sudden attack by the Danes. So, in 857, he built his wooden fort with drawbridges over the lake and the story of Leeds Castle began — a story dominated ever since by the fact that the Castle stands in the middle of a beautiful lake. At first the lake was regarded only as a water defence to be continuously improved by the Plantagenet military engineers and the English Mediaeval Kings. But from the beginning the Castle must have had a romantic beauty and, by a delightful paradox, this has proved more potent in preserving the Castle down the ages than its thick stone walls and drawbridges.

Eleanor of Castile and Edward I, passing by from Dover on their way back from the ninth Crusade, decided to buy it because of its beauty and seeming kinship with her father's castles in Spain. Froissart, the first European historian, writes in 1395 of visiting Richard II at 'a beautiful palace in Kent called Leeds Castle'. And Lady Baillie unquestionably bought it in 1926 because of what Philip Howard in *The Times* called 'all the attributes that old romantic Mallory or a Hollywood film director would expect in a castle . . . in the middle of a lake seeming to sail there like the black swans . . . a lake that you would expect an arm waving a sword to emerge from'.

So the meandering of the river Len on its way to the Medway and the Thames and the creative act of the Saxon rulers, Ethelbert and his Counsellor, Leed, have brought to this Kentish valley a cavalcade of notables down the ages. They have included Bishop Odo of Bayeux, who planted the vineyard and was then disgraced and exiled; Hamon de Crevecoeur, the Norman Baron, cousin of William the Conqueror, who fought at Hastings and whose son built the stone Castle in 1119; Lord William de Leyburn who, as Commander of the King's Fleet in 1294, became the first man in English history to bear the title of 'Admiral';

Edward I and Eleanor of Castile, who bought the Castle in 1278 and began some 300 years of Royal ownership; Henry Yevele, the famous architect who built the nave at Canterbury Cathedral and the roof of Westminster Hall; Sigismund, the Holy Roman Emperor; and Henry V, who negotiated a defensive treaty after the battle of Agincourt; Henry V's Queen, Catherine of Valois, who later fell in love with Owen Tudor and founded 'the greatest of all the Royal lines of England'; Henry VIII, who greatly enlarged the Castle and built the Maidens' Tower; Lord Culpeper, who was given 5 200 000 acres of Virginia by Charles II; Lord Fairfax, who emigrated to America and gave George Washington his first job; and Lady Baillie, who restored and cherished the Castle and, in Sir Arthur Bryant's words, with 'her vision and generosity, after a lapse of four centuries, has now brought about its re-dedication to the service of the nation'.

Alfred the Great's famous Bishop Asser wrote of King Ethelbert IV, whom Leed served: 'He governed with love and honour'. This is high praise and a moving memory from the golden age of the Saxon Kings, when Leeds Castle was born, and cannot fail to inspire the Trustees of the Leeds Castle Foundation as they strive to carry out the charitable duty bequeathed to them by Lady Baillie. And in this spirit we are proud that, with the invaluable guidance of Sir John Wilson, we first proposed to the British Government the holding of the Seminar on 'The Prevention of Disablement'. This led to the 'Leeds Castle Declaration on the Prevention of Disablement', now incorporated in the United Nations' 'World Programme of Action Concerning Disabled Persons' and we hope that this will in the years to come help millions of people, who would have been blind or maimed, to be healthy and happy.

The Leeds Castle Foundation would like to acknowledge the seminal role of Mrs Mary Lasker and Mrs Alice Fordyce of the sister Leeds Castle Foundation in the United States in proposing the relationship with Sir John Wilson, which led to the idea of the seminar on prevention, and also to thank Merck Sharpe & Dohme International, Barclay's Bank International, and the Leeds Castle Foundation (US) for their financial contributions towards the expenses of the seminar.

Contents

Leeds Castle Declaration on the Prevention of Disablement

12 November 1981

Towards the close of this International Year of Disabled Persons an international group was convened by the British Department of Health and Social Security at Leeds Castle, under the chairmanship of the former Prime Minister Lord Home of the Hirsel. It was a unique group of scientists, doctors, health administrators, and politicians, met to consider practical measures to prevent disablement, which afflicts one in ten of the human race. Their recommendations for immediate action to follow up effectively on the public concern stimulated by the Year of Disabled Persons were unanimous.

1. Disablement is a tragedy in terms of human suffering and frustration, and in terms of numbers. The number of disabled people in the world today is estimated at 450 million, of which one-third are children and four-fifths live in developing countries. Population growth and the increasing proportion of older people can only magnify the problem. Unless decisive action is taken now the number of disabled people could double by the end of the century.

2. Much of the underlying impairment is preventable. World-wide expansion of a programme of immunization could save five million children a year from disabilities caused by poliomyelitis, measles, tetanus, whooping cough, diphtheria, and, to a limited extent, tuberculosis. This could come about in ten years at a unit cost of about three dollars per immunized child.

A world-wide expanded programme of immunization would also facilitate production and effective use of vaccines against other diseases causing death or serious impairment. Among these is rubella, a prime cause of congenital blindness, deafness, and mental impairment. The use of rubella vaccine should be promoted in all countries.

3. Impairment arising from malnutrition, infection, and neglect could be prevented by inexpensive improvement in primary health care. Collectively these conditions now disturb the lives and reduce the productivity of at least twenty million people each year.

For example, trachoma and vitamin A deficiency blind at least two million people annually; this can be controlled. Similarly, inexpensive and simple treatment can arrest impairment from leprosy (afflicting at least three million people), can restore sight to ten million people blinded by cataract, and can improve the hearing of ten million deaf people.

Particular attention should be paid to the nutrition of pregnant women to prevent malnutrition of the fetus and to encourage breast-feeding of the baby.

4. There are many opportunities for improvement in regard to other disabilities. These depend on more effective sharing of knowledge, especially with the public.

Many disabilities of later life can be postponed or averted. There are promising lines of research for the control of hereditary and degenerative conditions. Early identification and treatment of raised blood pressure could save millions from premature disability and death due to heart disease and stroke. The toll of accidents and addiction could be markedly reduced.

5. Disability need not give rise to handicap. Failure to apply simple remedies very often increases disability, and the attitudes and institutional arrangements of society increase the chance of disability placing people at a disadvantage. Sustained education of the public and of professionals is urgently needed.

6. Avoidable disability is a prime cause of economic waste and human deprivation in all countries, industrialized and developing. This loss can be reduced rapidly.

The technology which will prevent or control most disablement is available and is improving. What is needed is commitment by society to overcome the problems. The priority of existing national and international health programmes must be shifted to ensure the dissemination of knowledge and technology. With proper use of modern communications this would involve modest costs and would bring great economic benefits. For instance, the world community is saving itself one billion dollars per year by the eradication of smallpox.

7. Although technology for preventive and remedial control of most disabilities exists, the remarkable recent progress in biomedical research promises revolutionary new tools which could greatly strengthen all interventions. Both basic and applied research deserve support over the coming years.

8. The members of the Seminar gave full support to all aspects of the programme of work contained in the Year of Disabled Persons, and in particular welcome the success of the disabled people themselves in bringing their frustrations and ambitions before the global community. It is only within individual communities and families that the lot of the disabled can be properly improved. Mobilization of the political will to act is a necessary basis for any successful programme.

9. In the view of this international Seminar, a programme of action to prevent disablement is a logical and essential part of the follow-up to the International Year of Disabled Persons. It would ensure that the next generation did not suffer from the present degree of avoidable disability, and would constitute a most appropriate, effective and long-lasting contribution to the health and happiness of mankind.

Participants

The Rt. Hon. The Lord Home of the Hirsel — *Chairman.*

A.R. Al-Awadi, Minister of Health for Kuwait.

H.B. Betts, Executive Vice President and Medical Director, The Rehabilitation Institute of Chicago.

G.A. Brown, Deputy Administrator, United Nations Development Programme.

S.G. Browne, Secretary, International Leprosy Association.

W.D. Clark, President, International Institute for Environment and Development.

G. Dybwad, Professor Emeritus of Human Development, Brandeis University.

G.J. Ebrahim, Reader in Tropical Child Health, University of London Institute of Child Health.

The Hon. Harry Fang, President, Rehabilitation International.

G. Foggon, Director, International Labour Office, London.

Alice Fordyce, Executive Vice President of Albert and Mary Lasker Foundation and Trustee of Leeds Castle Foundation (US).

D.A. Henderson, Dean, School of Hygiene and Public Health, The Johns Hopkins University.

D.E. Hyams, Senior Director — Medical Operations, Merck Sharp & Dohme International.

M. Irwin, Senior Adviser on Childhood Disabilities, UNICEF.

B. Jazbi, Federal Ministry of Health, Pakistan.

Lucy Kadzamira, Principal Nursing Officer, Ministry of Health, Malawi.

L.A. Kaprio, Regional Director for Europe, World Health Organization.

Y.P. Kapur, Professor of Otolaryngology, Michigan State University.

V. Ramalingaswami, Director-General, Indian Council of Medical Research.

B. Sankaran, Director, Division of Diagnostic, Therapeutic, and Rehabilitative Technology, World Health Organization.

E.C. Shore, Deputy Chief Medical Officer, Department of Health and Social Security.

J. Stokes, Medical Adviser to Leeds Castle Foundation Trustees.

L.B.J. Stuyt, Chairman, National Committee of the Netherlands for IYDP.

J.K. Thompson, Commonwealth Secretariat Consultant on IYDP — *Rapporteur.*

Sir John Wilson, President, International Agency for the Prevention of Blindness.

P.H.N. Wood, Director, ARC Epidemiology Research Unit, University of Manchester.

1

The scale of the problem

During the International Year of Disabled Persons, 1981, there was general acceptance of a figure between 450 000 000 and 500 000 000 for the number of disabled people in the world. This estimate owed its origin to earlier surveys conducted by Rehabilitation International. Subsequent calculations by other institutions, including the World Health Organization, confirmed that the total is probably within this range. UNICEF adopted the same estimate ('One in Ten') adding that four-fifths of the total live in the developing countries and that, of those, about a third are children.

Such figures present a challenge to the world's conscience. As a basis for action they would have greater influence if, nationally as well as internationally, more precise definition were given to the concept of disability and if disabilities were considered in their component parts and their geographical distribution. Physical, mental, or sensory impairments may be of short or long duration; and they vary widely in their degree of severity. In the absence of adequate statistics to enable numbers to be assigned to short-term, long-term, or permanently disabled people or to the various degrees of severity, criteria are used relating to ability to 'function as fully as persons of their age and sex are expected to do in their societies', i.e. the degree of handicap resulting from the disability.

The concept of 'One in Ten' has become current and inevitably it is being applied as though it relates to the proportion of disabled people in every community throughout the world. During the International Year, many countries initiated surveys to establish more accurate figures for the numbers of disabled people. Where preliminary results have been collated, there is no reason so far to revise the global estimate; but better figures will in due course make it possible to analyse more clearly the categories of disablement, the main causes and the distribution between industrialized and developing countries and between age groups.

In industrialized countries, disability is linked with aging. Taking blindness as an example, the number of blind people in every 100 000 of the general population in the United Kingdom — and similar figures generally apply to North America and Western Europe — is about 200. However, in the British population over 65 years of age, the prevalence is 2300 per 100 000, equal to the worst blindness rate in Bangladesh. Over 75 years of age in the United Kingdom, there is a spectacular increase to 7170 per 100 000, the figure equal to the prevalence in a community stricken with river blindness (onchocerciasis) in West Africa.

This dramatic increase with age is evident also in relation to deafness. Surveys in Britain have shown that over 30 per cent of people with serious hearing defects are over 65 years of age and that, over 75 years of age, the prevalence of deafness is 17 times more than in the general population. The age link with rheumatic and respiratory disorders is obvious. Mental handicap can noticeably increase with age. The risk of stroke and even of accident rises sharply in later years. The degree of disability increases with advancing age; and the elderly are one of the fastest growing subgroups in the population of Europe and North America. This is a race between the science that is extending the span of life and the science which can determine the quality of that life. It is a race which must be won because upon it depends the expectation and life-style of an aging population. In the slogan of the 1982 World Health Day, the aim must be to 'add life to years'.

Applying the same calculations to the developing countries, the prospect is alarming. During the next fifty years, the population of these countries is expected to treble. The number above age 55 could increase fivefold. It is these upper-age groups which are vulnerable not only to tropical and deficiency diseases but also to the age-linked disabilities which are at present associated primarily with industrialized countries. The growing number of disabled people is a problem which concerns every country at every level of development. To quote Mrs Gandhi at the May 1981 World Health Assembly, 'Humanity is one, at least in its vulnerability'.

A point which stands out in the chapters of this book is the inadequacy of the statistics which are available. Even the main causes of disability are not recorded in most countries. And there is little information about the age-levels of disabled people, particulary in developing countries. Some of the assumptions that are made could be invalid, for example that there are more blind people than deaf people. Surveys that have been completed suggest that the reverse is true, particularly in the younger age groups. Another assumption is that hypertension and stroke occur mainly in conditions of affluence and stress and are therefore more prevalent in urban and western communities. Whilst it is probably safe to assume that they are more prevalent in communities where people live longer, there is so far little significant evidence of differences in the incidence of hypertension and stroke in people of the same age groups in different countries.

Surveys inaugurated during the International Year of Disabled Persons will, in due course, fill in some of these gaps; but properly conducted surveys can cost a great deal of money which developing countries can ill afford. Many of them doubt, with reason, whether it is ethical to count disabilities without offering treatment or relief. Eye-camp techniques have shown that it may cost as much to count heads as to cure eyes. Undoubtedly, statistics do exist which are not being accumulated and analysed, for example in hospital records, in census returns, and in community accounts for purposes of relief and other social services.

When the World Health Organization endeavoured some years ago to establish more realistic estimates of the prevalence of blindness in the world, it found that by simply accumulating systematically the evidence which did exist but had not been co-ordinated, often even nationally, it was possible to collect fairly reliable figures and make reasonable projections. These were published in *Data on blindness* which has subsequently been the basis of planning for action and for calculations about the economic cost of blindness. A similar effort, using similar methods, might be expected to establish far more reliable figures than those that now exist about the extent and causes of disability in the industrialized countries and throughout the developing world. Such an effort would be worth making, so long as it does not degenerate into an action-postponing statistical exercise.

One source of a great deal of information about the prevalence of disabilities in rural areas of developing countries will be the field-testing, inaugurated in 1981, of the World Health Organization's experimental manual *Training the disabled in the community*. This suggests how disabilities can be detected in rural areas and how rehabilitation can safely and effectively be carried out by members of the family or of the community or by the disabled people themselves. There is considerable feedback during the field-testing about the numbers and categories of disabled people in countries as diverse as St. Lucia, Botswana, and Kerala State (India). While the first intention is to bring rehabilitation benefits to some of the 98 per cent of the people of developing countries who are at present denied them, there is also the intention to tackle measures of preventive action which WHO estimates could reduce the incidence of disability by at least 50 per cent. The simple, locally manufactured aids suggested by the manual will undoubtedly bring benefits to a high percentage of people with handicaps; but a proportion will need more sophisticated treatment and means will have to be found to provide it. This new initiative will throw a great deal of light on the nature and extent of disability in hitherto neglected areas and, incidentally, should indicate the main categories of impairments which could have been avoided in the first place.

Part of this problem of inadequate statistics is the lack of uniformity in defining the different levels of severity. The World Health Organization has published (1980) 'for trial purposes' an admirable International Classification of Impairments, Disabilities, and Handicaps which has already gained wide acceptance internationally. It contains a severity scale, but it reflects the degree to which an individual's activity performance is restricted rather than the severity of the underlying impairments. It therefore includes the impact of various forms of tertiary prevention, listed under the headings 'enhancement', 'supplementation', and 'substitution'. In other words, a disabled individual is categorized according to his activity performance after account is taken of aids, appliances, and assistance which permit this level of performance. The factor of severity is a matter of great importance. Unless there is a clear concept of the measure of severity, with or without aids, appliances or assistance, the understanding of the scale of the problem and therefore the priorities for action could lack reality.

Though the disabilities of affluence, addiction, and mechanization are important in relation to research options and the activation of political will which controls the flow of resources from the rich to the poor countries, the massive disability is, clearly, in the developing countries. It is among the poorest communities in the developing countries that poverty breeds disablement and disablement breeds poverty, a vicious circle that the poorer countries can least afford. These communities are the target of every rational development programme — under-privileged, under-served, under-nourished, at the bottom of every economic and social heap. Diseases long controlled elsewhere still flourish and bring with them not only death but life-long disability. To reach these communities, we must break through the easy abstraction of the bedrock of undifferentiated poverty and identify those vulnerable communities where disability is a major cause and component of that poverty. Just as, in the industrialized countries, disablement clusters around the upper age groups, so, in those vulnerable communities of the Fourth World, there is a prevalence of disabilities which is one of the main reasons why they are staying at the bottom of the economic and social pyramid and are cemented into that bedrock of poverty. This applies not only to those special communities where disabilities affect a major part of the population, such as areas of Asia afflicted by deaf-mutism or leprosy or the onchocerciasis regions of West Africa, but to the general level of disability in the poorest 5 per cent of the world's population. The papers in this book by Professor Ramalingaswami on Malnutrition and Dr Ebrahim on Perinatal neglect show how disabilities in these communities arise from very general causes such as malnutrition, lack of perinatal care, pollution, malpractice, and neglect; and these are so general that any proposal to do anything about the disabilities resulting from them is open to the charge that it is nothing less than an attempt to alter the world economic order and close the widening gap between the rich and the poor. However, it is possible to identify those disabilities against which intervention has a good hope of success long before some revolution in economic conditions improves the general pattern of life.

For example, xerophthalmia, the destroyer of childrens' eyes, is undoubtedly due to deprivation, poverty, malnutrition. However, successful action can be taken against it by identifying, within the broad spectrum of deprivation, one specific element: vitamin 'A' deficiency during the vulnerable first few months of life. The vitamin deficiency can be corrected by a simple and inexpensive intervention — thus saving children from blindness long before there is any material improvement in economic conditions of the community as a whole. Other interventions could readily be described which, together, would control a very considerable amount of the disablement resulting from deprivation. For example, surgery to improve hearing might well be simplified, as the cataract operation has been simplified, to the point where it can be delivered through mass treatment at village level. Thousands of children in the endemic goitre belt of the sub-Himalayan region of India and Nepal are born each year mentally retarded and deaf-mute for lack of a trace of iodine in their parents' diet. For the six major disease hazards of childhood in developing countries, the Expanded Programme on Immunization of WHO and UNICEF is already having its effect long before the elimination of the social and economic conditions which made the children vulnerable to them in the first place.

Targets for action

The task is therefore to identify not only the overall size of the problem but the specific targets for action in terms of avoidable disability. The objective is clear. It is to reduce, within a time scale of not more than twenty years, the burden of avoidable disability. The priority must be those communities where this burden is excessive and for those diseases and conditions in respect of which there already exists a potential for control which can be delivered at acceptable cost or for which a potential can be expected to develop as a result of practical research in the foreseeable future. This implies that priority should not be accorded to those conditions where there does not exist this potential for economical control or against those conditions which do not have a mass impact. It was the identification of these priorities which was a main task — and the main success — of the Leeds Castle seminar in November 1981 and the action that has followed it.

Discussion at the seminar was based on two broad concepts. The first is that the approach must be multidisciplinary, and in the language of international health, 'disease-oriented, not impairment-oriented'. Rubella is a good example of this. The attack must not be on the impairments — deafness, crippling, mental handicap, and blindness — which result from rubella but concerted action on the disease itself. It is not only the approach to the attack and treatment that must be multidisciplinary but also the approach to the training of the staff that will be required to implement national programmes. The question is whether this multidisciplinary approach can be institutionalized in the training of that staff — scientists, clinicians, planners, and administrators who are concerned not just with one disability but with disablement as a whole, whatever may be its cause. The second of the two broad concepts is that a separate, vertical programme of disability prevention

would be prohibitively expensive and politically unattainable; consequently what is required is a shift in the emphasis of a number of ongoing programmes so that they will include, as a distinct and important
objective, practical measures for the prevention of disability. The
attraction of this policy of integration instead of separateness is eco-
nomy and political feasibility. The danger is that the objective of disability prevention could easily be submerged under the broad flow of
more general programmes. Experience warns that the inevitable depart-
mentalization of the administration of such programmes will make
effective co-ordination difficult to achieve.

Within the multidisciplinary approach, there is a need for precise
objectives. It will be essential within each programme to define targets
which are accepted as part of the overall objective of the programme.
There is nothing new in this, because in world health planning and in
more general development programmes, the concept of precise targets
has become familiar, and there are sophisticated means of achieving it.
Co-ordination is likely to present more of a practical problem because
it will involve joint action by a number of United Nations agencies; the
Executive Board of WHO has proposed, as a result of the Leeds Castle
seminar, to appoint a consultative group to advise on practical, co-
ordinated measures for the prevention of disablement within ongoing
global programmes of health and development, and to monitor their
effects. At a United Nations inter-agency meeting in Vienna in March
1982 it was agreed that the prevention aspects of disablement should be
included as a major objective of the World Programme of Action concerning Disabled Persons.

Co-ordination at the national level may be even more difficult. Nearly
every country in the world appointed a National Committee for the
Year of Disabled Persons in 1981. A good deal of co-ordination between government departments and voluntary agencies, including associations of disabled people themselves, was accomplished by the work
of these committees. However, the achievements were mostly in the
fields of public awareness and in the education, employment, and
general well-being of the already disabled. Prevention aspects were in
most cases limited to the prevention of accidents on the roads, at work
and in the home, or the minimizing of handicap by the provision of
aids and appliances. In many countries, Councils for the Handicapped
have been appointed (or renewed) to continue the work of the National
Committees. There is growing awareness that, by co-ordinated action,
much of the disablement with which they are concerned could be
prevented.

What, then, are the global programmes which appear to offer an
immediate opportunity for co-ordinated action for the primary prevention of disability? The global concept of primary health care —
which took its origin from the Alma-Ata Declaration and which is now

the main thrust of World Health policy and of the aspiration of 'Health for All by the Year 2000' — offers an immense potential for controlling avoidable disabilities associated with malnutrition, infection, neglect, malpractice, and ignorance. A powerful weapon is the Expanded Programme on Immunization which already has targets which are specifically related to disability and which the Leeds Castle seminar hoped to see expanded to include immunization against rubella, meningitis, and, eventually perhaps, those aspects of tuberculosis which are not already covered and, one day, leprosy. To add disability prevention as a precise objective of primary health programmes, with an equal emphasis on morbidity as on mortality, would help to give such a programme that human dimension essential to popular support and motivation. The limiting factors are the capacity of the primary health worker, who already has so many tasks to perform, and the provision of supervisory responsibility from the village level to major health institutions.

The various child-care programmes are already beginning to have a conscious component for the prevention of disability. UNICEF estimates that in developing countries one-third of disabled people are children and as, increasingly, such children survive they project into the future a massive medical, social, and economic problem.

In terms of environmental control, the Water Decade could well be the most decisive advance of the 1980s. Even if the prevention of disability is not a major objective of the Water Decade, the supply of pure water and better sanitation will have a dramatic effect on the health of those at present denied them. The same can be said of programmes whose main objectives are the maintenance and improvement of food supplies and the promotion of better standards of nutrition.

The Leeds Castle seminar was devised to give a much-needed emphasis to primary prevention and, where necessary, to redress the balance in those areas where the International Year concerned itself so largely with public attitudes to people already disabled and their general well-being in the community. The seminar also identified a cluster of related and fairly inexpensive interventions at the secondary level of prevention, the arresting of the deleterious effects of otitis media, the cause of so much deafness, and relatively simple interventions to relieve the effects of crippling, whether from residual poliomyelitis, from accident, or from other cause. A programme of delivery for cataract operations has been well established in Asia. One question needing an answer is whether the Eye-camp technique can be used to bring to rural areas a package of simplified surgery and other forms of treatment for the relief or cure of a range of disabilities. The possibility needs to be explored of providing the conditions, as in an Eye-camp, where a variety of specialists could work on their particular surgical jobs within a rural framework that provides general back-up and administrative services, including anaesthetics, nursing follow-up, accommodation, and transport. This is now being successfully attempted for the orthopaedically disabled in India.

Developmental conditions, particularly those which result in mental disorder or retardation, must be a major concern of any programme for the prevention of disability. People who, in the words of Dr Philip Wood's subdivision on impairments, suffer from lack of awareness of self and surroundings may well constitute the largest group of the disabled both in industrialized countries and in developing countries. The difficulty, and it was noted in relation to social aspects of disability during the International Year, is the absence of reliable statistics about mental handicap and the problems which even the most expert authorities experience in defining terms and policy priorities.

Dr Dybwad, in his chapter on this subject, refers forcibly to the need for early recognition and stimulation and he considered that the Leeds Castle Declaration had been inadequate in its reference to mental disorder and retardation. The difficulty was that, though much has been written about action at the secondary and tertiary level and the need for radical change in social attitudes, surprisingly little authoritative information is available about primary prevention in relation to mental handicap.

The specialist organizations, and all who are concerned with the comprehensiveness of action against avoidable disability, should give increased attention to this subject.* Areas of study should include prenatal care, protection of the fetus from infection, malnutrition, and deleterious drugs. Increased research is also needed into complex heredity factors and into those environmental interventions which might be practicable and acceptable, e.g. to control lead pollution or to increase iodine intake in areas where goitre is endemic.

In both industrialized and developing countries, injury from accidents in the home, on the roads, and at work are major causes of disablement. Dr Kaprio's paper on injury prevention shows that, because of improving general levels of health in industrialized countries, the relative importance of accidents as a cause of death and injury is increasing, particularly among the young and the elderly. Mr Foggon's paper on occupational accidents points out that increasing industrialization in

---

*A report dated January 1980 (typescript) by the Joint Commission on International Aspects of Mental Retardation, commissioned by the World Health Organization indicates that the specialist organizations are already giving it a great deal of concerted attention. The report contains a major section on primary prevention and concludes: 'There has been a revolution in our knowledge of the prospects for prevention and amelioration, a revolution which now makes it possible to diminish significantly the impact of mental retardation on the individual, his family and society at large'. It is not yet clear what action will be taken internationally on the detailed action proposals in this document. But it is significant, in the context of this book, that the report comments: '. . . the gap between our existing knowledge and its application remains too wide'.

developing countries, including the mechanization of agriculture, exposes the workforce to new risks and dangers without adequate protection. Measurable success has been achieved in many countries in the control of traffic and industrial accidents. It is for each country to decide what action it will take in regard to such matters as the compulsory wearing of seatbelts and the enforcement under penalty of internationally agreed standards of safety at work. However, even in communities which traditionally cherish the liberty of the individual, the question may reasonably be asked whether that individual, through negligent use of the lethal resources now at his disposal, has an inalienable right to injure other individuals or himself. As an authority on traffic accidents said at the seminar, 'Your death may cost me nothing, but your paraplegia could cost me and my fellow tax-payers a million dollars'.

2

The principal causes – developmental conditions

Vaccine – preventable disease

The role of the immunization services

Ralph H. Henderson *

Immunization is one of our most powerful and cost-effective weapons of disease prevention, yet remains tragically under utilized. In the developing world today, 10 children die and another 10 become disabled with each passing minute because of the lack of availability of immunizations. Diseases such as neonatal tetanus and poliomyelitis, which have been virtually eliminated in most of the developed world, continue to take a horrible toll in the developing world. Measles, which kills only some two of each 10 000 cases in the United States, kills two per 100 cases in the developing world, the figure rising to 10 or more in malnourished populations. Lacking immunization, almost all children will contract this disease. Whooping cough is another major killer, particularly in the child of less than six months of age, and diphtheria and tuberculosis can be added to the list as additional important problems. During the course of a year, five million children will die and another five million will become crippled, deaf, blind, or mentally retarded because of these six vaccine-preventable diseases.

The high incidence and great severity of these diseases in the developing world is due to a vicious cycle of infection and malnutrition, with either factor capable of initiating the cycle. Children in developing countries often have their defence mechanisms compromised from the start by low birth weight and then are assailed by a series of stresses which include whooping cough, measles, and weaning on top of repeated episodes of diarrhoea and malaria. Each event sets the child back in growth and development and, if the interval between events is too short, the child spirals down to death.

Immunization services are effective in preventing specific diseases which can precipitate malnutrition and, by permitting the child a longer recovery period between the events mentioned above, can prevent this downward spiral and contribute significantly to the overall reduction of childhood mortality.

By preventing some of these events, immunization services can help reverse this cycle, and their contribution to the prevention of infant and childhood disability and death can therefore extend beyond the prevention of the individual target diseases. But immunization services

*Director, Expanded Programme on Immunization, World Health Organization, Geneva.

are themselves best delivered along with other services of high relevance to pregnant women and children in their first year of life, the persons who represent the prime concerns of immunization programmes in the developing world. These services include oral rehydration for diarrhoeal diseases, malaria treatment and prophylaxis, and counselling with respect to nutrition during pregnancy, breast-feeding, weaning, clean water, and sanitation. Not only do such services act in synergy to break the vicious cycle of infection and malnutrition, but each service helps promote the utilization of the other services by the population concerned, leading to greater efficacy and lowered cost per service delivered. Thus the effort to make immunization services available in the developing world should be viewed as an integral part of the effort to strengthen preventive and curative services available to mothers and children.

It is in recognition of the above that immunization has been cited as one of the essential elements of primary health care in the Declaration of Alma Ata, and that WHO, with the help of UNICEF, UNDP, national donor agencies, and voluntary agencies, is sponsoring the global Expanded Programme on Immunization whose goal is to provide immunizations for all children of the world by 1990. A good start has already been made, with some 139 developing and developed countries implementing this programme, and with over 2000 professional national and international staff trained in EPI management courses. It is estimated that in 1981, US $ 72 million was invested in the programme, two-thirds from the budgets of the developing countries themselves. But at an estimated cost of US $ 3.00 per fully immunized child, some US $ 300 million will be required each year by the end of the decade to reach the 100 million children needing coverage in the developing world. So, although much has been done, much more is required.

The challenge is great, but the promise is even greater: for a cost amounting to only US $ 0.12 per capita in the developing world, the six killer scourges of diphtheria, measles, poliomyelitis, tetanus, tuberculosis, and whooping cough can be eliminated as public-health problems of the young child. The elimination of these diseases will constitute a major contribution to the prevention of infant and childhood disability and death, and this, by demonstrating to parents that their children are likely to live, will diminish the incentives for having many children. Furthermore, by establishing vaccine-delivery systems in the developing world which are capable of achieving high coverage of susceptible populations with vaccines known to be safe and potent at the time of use, the stage is set for the introduction of other vaccines, particularly new vaccines now being developed through research, which will increase the overall effectiveness of immunization services. This is a challenge which must not be refused!

A FIVE-POINT ACTION PROGRAMME FOR THE 1980s

The WHO Global Advisory Group on the Expanded Programme of Immunization has met annually since 1978. During its meeting in October 1981, it reviewed a draft progress report and concluded that much progress had been achieved. But it also recalled that because immunization services in the developing world are still not generally available, 10 children die and another 10 children become disabled with each passing minute. It warned that the current rate of programme progress was not sufficient to achieve the EPI goal of reaching all children by 1990, representing not only a setback for EPI but also a threat to WHO's aspirations for achieving health for all by the year 2000. Reaffirmation of national commitment and intensification of programme activities are needed, and the Global Advisory Group endorsed the following five-point action programme as a guide for national and international efforts for the remainder of the decade:

1. Promote EPI within the context of primary health care

☐ Develop mechanisms to enable the community to participate as an active partner in programme planning, implementation, and evaluation, providing the technical and logistical resources to support these functions.

☐ Deliver immunization services with other health services, particularly those directed toward mothers and children, so they are mutually supportive.

2. Invest adequate human resources in EPI

Lack of these resources in general, and lack of management and supervisory skills in particular, represent the programme's most severe constraint. Capable senior and middle-level managers must be designated and given authority and responsibility to carry out their tasks. They require training, not only to be effective with respect to EPI, but also to contribute to the understanding and strengthening of the primary health-care approach. Reasons for low motivation and performance in the areas of field supervision and management need to be identified in order to take appropriate measures to encourage managers to visit, train, motivate, and monitor the performance of those for whom they are responsible.

3. Invest adequate financial resources in EPI

For the programme to expand to reach its targets, current levels of investment in EPI, estimated now at US $ 72 million per year, must be doubled by 1983 and doubled again by 1990 when a total of some US $ 300 million (at their 1980 value) will be required annually. Over two-thirds of these amounts must come from within the developing countries themselves, the remaining one-third from the international community.

4. Ensure that programmes are continuously evaluated and adapted so as to achieve high immunization coverage and maximum reduction in target disease deaths and cases

Such adaptation depends on the development of adequate information and evaluation systems. By the end of 1985 at the latest, each country should be able to:

☐ estimate reliably immunization coverage of children by the age of 12 months with vaccines included in the national programme;

☐ obtain timely and representative reports on the incidence of EPI target diseases included within the national programme; and

☐ obtain information on the quality of vaccine so that it is known that the vaccines employed for EPI meet WHO requirements and are potent at the time of use.

In addition, countries should promote the use of periodic programme reviews by multidisciplinary teams comprised of national and outside staff to ensure that operational problems are identified and that a wide range of experience is reflected in the recommendations which are made.

5. Pursue research efforts as a part of programme operations

The objectives should be to improve the effectiveness of immunization services while reducing their costs and to ensure the adequate supply and quality of vaccines. Specific concerns include the development of approaches for delivery services which engage the full support of the community, the improvement of methods and materials relating to sterilization and the cold chain, the acquisition of additional knowledge concerning the epidemiology of the target diseases, further development of appropriate management information systems, and further improvement in the production and quality control of vaccines which are safe, effective, and stable.

The Global Advisory Group believes that, if countries accept the achievements of EPI as a principal indicator of the success of their strategy to achieve health for all by the year 2000 and if they will address current programme constraints through the action programme, the goal of providing immunization for all children of the world can be attained by 1990.

Is Eradication of poliomyelitis possible?

*L. B. J. Stuyt**

The National Committee for the Year of Disabled Persons, 1981, in The Netherlands, aware of the great importance of the promotion of preventive measures to avoid disablement, followed the recommendations

*Chairman, National Committee of the Year of Disabled Persons in The Netherlands.

of the General Assembly of the United Nations to assist developing countries most afflicted by disabling diseases such as poliomyelitis by promoting vaccination against these diseases. Successful immunization of the population would greatly reduce the need for rehabilitation measures. It is understandable that the question has been raised during the year 1981 why eradication of poliomyelitis could not be started world-wide, as has been done with smallpox.

It is now generally accepted that the eradication of smallpox virus as a pathogen of man has been successful. This success was due to a scientifically sound epidemiological analysis of the transmission of the virus infection, i.e. the emphasis was laid on case-finding followed by ring-immunization in addition to the traditional inclusion of smallpox vaccine in a general immunization programme.

The lessons derived from the WHO smallpox eradication programme should be kept in mind in starting the implementation of the Expanded Programme on Immunization of the World Health Organization (EPI) now being developed.

The two basic problems of this programme are:

1. How can the best immunization be developed comprising vaccination against six or seven diseases (poliomyelitis, diphtheria, tetanus, pertussis, tuberculosis, measles, and yellow fever) for use in a wide variation of different cultural, political, and economic situations?

2. How can the availability of sufficient amounts of vaccines of good quality, and at low price, be ensured for this purpose?

It will be clear that necessary conditions for successful immunization are: a stable, efficient vaccine, available in great quantities and at economically acceptable prices; sufficient well-trained personnel for the administration of the vaccines; and finally sufficient collaboration and willingness on the part of the population to be vaccinated.

Until recently poliomyelitis vaccination formed a weak link in the chain of vaccinations used in developing countries. There exist two types of poliomyelitis vaccine:

1. The oral poliomyelitis vaccine, OPV, originally developed by Sabin. It is given as a mixture of the three existing attenuated (avirulent) vaccine strains. In order to be effective, i.e. to induce a lasting immunity against the disease, the live vaccine has to multiply in the gut; this has caused difficulties, especially in tropical areas. In addition the vaccine is difficult to transport in tropical countries, as it has to be kept at $-20°$ C. Oral administration to babies is not always easy. However, in countries with a moderate climate it has been very successful.

2. The inactivated poliomyelitis vaccine, IPV, developed by Salk, which is administered by injection. Some countries, among them The Netherlands, decided some 20 years ago to use this type of vaccine in their immunization programmes. To produce this type of vaccine, virulent virus is multiplied on living monkey-kidney cells, which can

only multiply as a thin layer on glass or plastic surfaces. After multiplication the poliovirus is inactivated, and the effectiveness of the vaccine is controlled by its property to induce antibodies in laboratory animals.

Scientists in the National Institute of Health in the Netherlands (Rijksinstituut voor de Volksgezondheid, RIV), responsible for the production and control of this vaccine, realized 20 years ago that they could hope to continue the use of this IPV only if its quality could be further improved. Originally, it induced persisting immunity in 'only' 90 per cent of the immunized children, using three or four injections. The approach made by them to improve this situation contained a number of interesting technical steps, two of which will be discussed briefly.

1. The first step was to replace the two-dimensional system of the cell cultures, grown on glass surface, by a three-dimensional one, with the aim of multiplying the virus in much greater quantities in large tanks instead of on the surface of relatively small glass bottles. As they realized that the cells could never be induced to multiply in suspension, small plastic beads, called microcarriers, with a diameter of 0.1–0.2 mm, were used in order to provide the kidney cells with a large surface for multiplication. It was then possible to produce much larger quantities of virus in 350-litre stainless-steel fermentors. After concentration and purification, a potent inactivated poliomyelitis vaccine could be produced in large quantities.

2. The second achievement, following as a result of the first, was the reduction of the number of monkeys needed for the procurement of kidney cells. Originally large quantities of monkeys had to be imported, which became more and more difficult. By improving the procedure to isolate individual cells from the kidney, and by using secondary and tertiary cells, by allowing the cells more divisions, the cell yield has greatly increased. The number of monkeys needed for the production of 1 000 000 doses of vaccine for human use has now been reduced from 1000 to five. Moreover it was found to be feasible to breed the small number of monkeys still required in captivity, so that importation is no longer needed.

By applying these and other technical improvements it was possible to produce finally an IPV with a known and constant content of virus antigen of the three existing types of virus. Such vaccines have now been tested thoroughly in different countries in the developed, as well as in the developing, world. On the basis of the results of some carefully executed trials it has been possible to determine optimal amounts of active material (antigens) for the three polio-virus types in the highly improved vaccine. It has become obvious that all vaccinated children develop after the first injection an immunity of at least 80 per cent whereas after the second injection, given six months later, the immunity rises to almost 100 per cent. Antibodies continue to exist in the blood for many years, and experts now consider that the immunity after two injections will be of a life-long nature.

When it became clear that life-long immunity against poliomyelitis could be obtained with vaccination in two sessions with an interval of six months, ways were sought to adapt the existing four-session immunization schedule against D (diphtheria), P (pertussis), T (tetanus), and P (poliomyelitis) into a two-dose schedule.

In the last two years French experts of the Association pour la Promotion de la Médecine Préventive (APMP) working in Senegal (Kolda zone) have successfully applied an immunization schedule, carried out in two sessions, using as a standard procedure a DPT–polio vaccine, administered with a six months interval. BCG vaccine was given simultaneously with the first dose of DPT–polio vaccine at the age of 3–8 months, while measles and yellow fever vaccines were administered in the second session, simultaneously with the second DPT–polio vaccine injection (age 9–14 months). It has become clear now, from carefully executed controls, that poliomyelitis and measles have disappeared from the area.

On the initiative of the National Committee of the Year of Disabled Persons 1981 in the Netherlands, a similar scheme of vaccination will be introduced in Upper Volta, in the province of Kaya. In full agreement with the Health Authorities of that country, the whole area will be vaccinated with the two-dose scheme, with an interval of six months. A careful evaluation will be conducted after the first year.

Although the two-dose scheme with a six-months interval, and the better quality of the vaccine, have tremendously improved the immunization technique and results, all the problems have not yet been solved. The most important remaining problem is the factor of *cost* to produce the poliomyelitis component of the DTP–polio vaccine. The experts have several approaches to solve this problem. In the first place, the new technology with newer cell-cultures, as described above, increased production at lower cost. In the second place, a more industrial technology with 10 000-litre steel containers will lower the cost even more. In the third place, a vaccine-producing laboratory working in close collaboration with the RIV has been opened in Egypt. Producing these vaccines in Africa will undoubtedly lower the costs for developing countries in that area.

Finally the plasmid carrying the genetic code for the production of type I polio antigen has recently been made available by researchers of MIT to their colleagues in the RIV. Dutch scientists are now trying to bring it to expression in a bacterial strain, which is generally used throughout the world for this type of recombinant DNA work. If this approach is successful the yield of active materials to be used in a polio vaccine might well be increased at least tenfold.

Thus many factors work together to increase production, improve quality, and lower the costs.

The experts working in this field dream about an ultimate purpose of their studies: eradication of poliomyelitis in the whole world. It cannot be claimed that this is yet feasible, but its realization seems now much nearer than was expected a few years ago.

Acknowledgement

The author thanks Dr H Cohen, Director-General of RIV for his support in preparing this text.

Perinatal neglect

G. J. Ebrahim *

The most striking characteristics of maternal and child health services in the Third World are their poor state of development, the urban bias, the emphasis on curative care utilizing complex technology and the focus on prestigious institutions. The most widespread and pernicious health problem is the lack of coverage of the population with basic health care. A joint UNICEF/WHO study on the state of health services in 1976 mentions that not more than 20 per cent of the rural population in the average developing country receive basic health care on a regular basis. The state of the urban poor is no better in spite of the relative proximity of hospitals and medical institutions.

This lack of regular health care has led to a huge backlog of unmet needs. Thus, in any given year there are about 100 million children suffering from malnutrition with approximately two million dying from it. Five million child deaths occur annually from diarrhoeal disease and another five million from preventable infections like measles, whooping cough, and diphtheria. The figures for morbidity from these illnesses and parasitic diseases of the tropics are even higher. These diseases are all amenable to prevention. Commonly described as the diseases of poverty they are truly speaking diseases of neglect since they arise when basic health needs are not met. Their contribution to disability in later childhood or adult life has not been measured or indeed even observed. However, since disease and disability go together malnutrition and recurrent infection, with one perpetuating the other and occurring in an environment of poverty, are the root causes of disability.

The process of child growth and development is complex with continuing interaction between the genetic potential and the environment. The result of this interaction at each stage sets the pattern for the next one. Recent studies on child development help us to derive several conclusions viz.

*Institute of Child Health, University of London.

☐ Any damage either through deficiency or an environmental agent in the formative years of early childhood can have serious influence on later development. The earlier the age at which such damage occurs the more serious the effects and the poorer are the chances of full recovery.
☐ Recovery from acute short-lived stress can be complete. It is when stress is chronic and especially multiple that permanent damage is likely.
☐ The home environment and the family's ability to provide nurturing care with adequate stimulation is crucial in the ability of children to live with a handicap.

Malnutrition and the health of mothers and children

The effects of malnutrition are most serious in early life and especially during fetal life. Inadequate nutrition in fetal life is a common cause of low birth weight defined as weight at birth of less than 2500 g. We know that there is a higher incidence (four to six times normal) of physical and mental handicap in infants of low birth weight. Mortality in the newborn period is also eight to ten times that in infants of adequate weight, and this increased likelihood of death is present up to the age of one year. Of the 22 million infants born each year with a low birth weight, 21 million are in the developing world. The majority of disabilities and infant deaths stem from this group. Several factors peculiar to the developing world are responsible for this high prevalence of fetal malnutrition. These are:

1. Marriages in most peasant societies occur early, usually around the age of menarche, when many mothers may not have completed their own growth. The first two years after menarche are specially hazardous since the chances of delivering low-birth-weight infants are doubled.

2. Many mothers may have suffered undernutrition and ill health in childhood. This is particularly so in societies where the status of the woman is low and boys are preferred over girls. The average adult female in the developing world thus tends to be shorter and lighter than her counterpart in the affluent societies of the West.

3. In communities with marginal nutrition the diet in pregnancy is often inadequate. Weight gain in pregnancy is almost half of that in affluent societies (4–6 kg as compared to 10–12 kg). The disturbing fact is that in many countries over the last two decades there have been virtually no improvements either in pre-pregnancy weights, food intake during pregnancy, or in average weight gain in pregnancy.

4. Pregnancies are often not adequately spaced so that the mother is not able to recover sufficiently from the nutritional cost of the preceding pregnancy.

5. Reproduction continues into middle-age and this may be one factor in chromosomal anomalies like Down syndrome. In Sweden as fertility rates among older women have declined the number of children with

Down syndrome has also decreased so that the incidence of mental retardation is reduced from five per 1000 in 1959-1962 to about three in 1967-70.

The growing spectre of marasmus hanging over the cities of the developing world

In the squatter settlements of the developing world bottle-feeding is on the increase with a consequent increase in malnutrition in early infancy. Hitherto malnutrition below the age of six months used to be a rare phenomenon. Now it has become common wherever bottle feeding has spread. Two further developments tend to aggravate the situation viz. the use of sweetened condensed milk for infant feeding instead of a proper powdered formula and the delegation of feeding to an older sib by the working mother so that the constitution of the feed can be dangerously faulty. Diarrhoea is the usual accompaniment of bottle-feeding. In early infancy as in fetal life the brain is growing maximally and is therefore most vulnerable to nutritional insults as well as to the electrolyte disturbance of diarrhoea. A further twist of this tragic situation has been added by the present trend of rapid urban growth so that by the year 2000 almost 40 per cent of the world's children are expected to be living in large urban sprawls and agglomerates. Unless services expand rapidly many of the world's cities will have on their hands a growing number of under-class citizens who have suffered stunting in mental and physical development.

The new epidemics of urban life

The automobile has become the great killer and maimer in all cities, but whereas the cities of the developed world have been able to contain the menace this is not so in the developing world. Environmental pollution with lead from exhaust fumes creates an additional hazard. In the developing world several traditional practices requiring the use of a black pigment exposes children and even the fetus to the danger of lead poisoning.

Childhood infectious illnesses

Measles, whooping cough, poliomyelitis, and diarrhoeal diseases are all severe illnesses. Their contribution to the etiology of malnutrition is only now being recognized. Secondly it is not yet realized that in developing countries they occur at an earlier age than as described in Western textbooks and hence the danger is greater. For example, in Guatemala 19 per cent of all cases of measles occurred in the first year of life and 22 per cent in the second year. In Kenya, 15 per cent of all cases occurred under the age of one year. For some of these illnesses their epidemic nature is only now coming to light. Data from 71 developing countries reveal that the incidence of poliomyelitis more or less trebled during 1961–65 in the case of at least 45 of them. Surveys of lameness in Ghanaian schools and in India suggest that the mean annual incidence rates of poliomyelitis in tropical endemic areas have always been as great, if not greater, as those experienced by temperate countries during epidemic periods in the fifties.

Role of maternal and child health services

There is sufficient evidence to prove that improvements in the health of mothers and children in the Western World have occurred as a result of national programmes for their protection and for ensuring that basic human needs are met. The two basic service programmes in Mother and Child Health are the Antenatal and the Under Fives' Clinics with emphasis on health surveillance, education, immunization, child-spacing, and the early selection of those at-risk. The key issue here is of coverage. Unless 80 per cent or more of the population are regularly covered by these services the benefits will be marginal. Secondly, these services need to be slanted towards the disadvantaged and towards those who feel unwanted or have become socially alienated. It is here that the major challenges lie. It is in these groups that illness and disabilities are likely to arise.

Amongst all the surveillance and promotive programmes the MCH services are best suited to become family and community based and to establish their roots in the social and cultural milieu of the community. It is only through such a base that the MCH services can begin to grapple with the so-called clustering phenomenon wherein disease, malnutrition, and disability tend to cluster in certain groups and families adding to their social disadvantage. In the developing world, as elsewhere, the person with the disability is at the greatest risk of disadvantage. Without supportive programmes the slow drift to the bottom of the pile is inevitable. The disabled constitute a large pro-portion of the world's poor. MCH services with a purely clinical approach will not get to the roots of these problems. What is needed is a new breed of health workers with the ability to identify and reach out to the disadvantaged in the community, to seek out and prevent disease and malnutrition and the resourcefulness of preventing these from causing disabilities.

Malnutrition

*V. Ramalingaswami**

The problems — their nature and dimension

THE QUARTET OF MALNUTRITION

Malnutrition itself is an all-pervading disability. It constitutes an important backdrop to tropical health and development.

There are four major nutritional disorders of the developing world which can lead to widespread impairment and disability. The first and most widespread and most intractable of all is *protein energy malnutrition* (PEM) which exerts deleterious effects upon physical growth and growth of the brain, immunological competence, ability to cope with environmental infections, and ability to heal wounds. PEM is a part of the wider problem of poverty and under-development. Between 2 and 4 per cent of children below the age of five, living in rural areas in India, suffer from severe forms of PEM. Nearly 60 per cent of children below the age of five suffer from mild to moderate degrees of PEM as judged by growth failure. The average birth weight of infants born to poor mothers in India is 2.8 kg in comparison with 3.2 kg in the well-to-do classes. Almost one-third of children in the poor income groups are born with a birth weight below 2.5 kg. At the age of one year, a child from the low-income groups in India weights 7 kg against 10 kg amongst well-to-do families.

PEM is a multifactorial problem and has to be tackled in the long turn along the lines of promoting primary health care and human development. There is a silver lining. Even those children that had advanced to the severe stage of PEM can still be rehabilitated back and the bulk of the deficits and impairments can be reversed. There are some doubts as to whether intellectual impairment can be reversed completely and also whether impairment of immunological competence acquired in intrauterine life can be reversed completely by post-natal feeding.

The second problem is *endemic goitre*. In 1960, the estimate was that there were 200 million persons across the world who suffered from this condition and there is little evidence for a drastic reduction in that number since then. The latest estimates indicate that in India alone, 40 million persons are affected with this condition. Severe endemic goitre is associated with increased prevalence of developmental abnormalities in the form of deaf-mutism and cretinism. Recent studies in the sub-Himalayan region of the State of Uttar Pradesh India using radio-immunoassay of T_3, T_4, and TSH revealed that up to 2 per cent of the newborn show convincing evidence of sub-thyroidism (N. Kochupillai,

*Director-General, Indian Council of Medical Research, New Delhi-110029, India.

personal communication). In the inner valleys of the Himalayas in Nepal up to 10–15 per cent of the population may be suffering from deaf-mutism, cretinism, and other forms of severe developmental disturbances.

Endemic goitre is the cheapest and easiest of all diseases to prevent. No more new knowledge is required. The provision of iodized salt so as to ensure a regular intake of as small a quantity as 100 μg of iodine every day will eliminate endemic goitre. The injection of iodized oil or its oral administration to pregnant women will wipe out deaf-mutism and cretinism. The iodized oil injections need to be administered only once in three to five years.

The third nutritional condition is *xerophthalmia* leading to nutritional blindness. Against a daily requirement of 300 μg of vitamin A, pre-school children in areas endemic for xerophthalmia consume, on the average, less than 100 μg of vitamin A. It is a matter of consuming a mere 40 g of green leafy vegetables a day that stands between the child and the protection of his eyes against blinding malnutrition.

The fourth nutritional condition is *nutritional anaemia*: 30–40 per cent of pregnant women and pre-school children in India and several countries of the developing world suffer from sub-normal levels of haemoglobin attributable to a deficiency of one of the cheapest elements on this planet, viz. iron. We are still unable to control this condition effectively.

Diarrhoeal disease is a major cause of mortality and morbidity in pre-school children. In several developing countries up to 30 per cent of beds in the children's wards of hospitals are occupied by cases of severe diarrhoea. Oral rehydration with glucose salt mixture is a major advance in reducing mortality from this condition. The easy availability of this simple technology and its early use following village-based training can bring mortality from this condition tumbling down.

Lathyrism is a crippling disorder resulting in irreversible paralysis of both legs and resulting from eating the toxic pulse, *Lathyrus sativus*. The disease attacks young men in their twenties. In some of the rural areas of the State of Madhya Pradesh in India up to 4 per cent of the population may be afflicted with this disorder. Through progressive stages those that are afflicted are reduced to crawling and become social outcasts. The economic implications of this widespread incapicitation can only be imagined. Several alternative approaches are possible — banning the cultivation of this pulse, propagation of low-toxin strains of the pulse, and parboiling.

Endemic fluorosis is yet another crippling disability affecting vast populations living in areas where the drinking water has high levels of fluoride. Fluoride-induced damage affects teeth in the form of dental mottling and bones with the result that deformities and paralysis set in. Alternative technologies have been developed for reducing the intake of fluoride to less than toxic levels in the home and community settings. The provision of alternate sources of water with lower fluoride content provides a lasting solution.

The challenges

The first challenge, as I see it, in tackling these problems, is the translation of scientific knowledge which is a perception of facts to public policy which is a system of values.[1]

When science and politics relate to one another effectively, society derives maximum benefit.[2] When public policies encourage research and also seek to apply the fruits of that research for human welfare, then progress ensues. The coupling of knowledge to action is the crux of the problem of development as a whole and of health development in particular.

The second challenge relates to the mismatch between our ability to grow food and our ability to see that it is accessible to all those in need. Today we stand disillusioned in this regard.

The third challenge is that the conquest of malnutrition requires co-ordinated multisectoral action. Over the years I have come to realize the importance of the human factor, that there is hardly any health condition in which pure hardware technology provides the complete answer in developing countries. Perhaps, we have been looking at the nutritional problem too minutely in a technological frame. Low income, low education, environmental insanitation, diminished food intake, repeated infections, family instability, low personal education, too many births, too closely spaced, all this we know constitutes the web or the face of poverty and under-development. It has to be tackled on a multiple front through an integrated package of technologies.

The approaches

We have had a large number of small-scale experiments and pilot innovations which have been successful in extending technologies and enlarging the outreach of services to rural areas. We need to make a big jump from such small-scale endeavours to mass application of existing and new devices in a new form of organization. In this process, community participation and village-level approach must receive focus. The greatest safeguard against malnutrition in infants and young children lies in the development of the insights of the mothers. Maternal caretaking, breast-feeding, the correction of faulty feeding practices during the first years of life, the deployment of community health workers and mothers themselves for health intervention in a primary health care approach with support from the professional personnel, efficient surveillance, these constitute a profile of approaches that can usher in the era of primary health care.

Experiments across the developing world clearly indicate that infants and child mortality rate can be reduced to one-third or one-half within five years at a cost less than the equivalent of 2 per cent of the per capita income.[3]

There can be a *health revolution* if society so wishes it just as there had been a *green revolution*. The time has come, as Dr Malher said, to move away from the timid gropings of the past and move towards bold large-scale interventions through a grand symphonic movement of primary health care. Nutritional disabilities will then pale into insignificance. They are a needless blot on society.

References

1. Stewart, W.H. The use of Government to protect and promote the health of people through nutrition. *Fedn Proc. Fedn Am. Socs exp. Biol.* **38**, 2557–9 (1979).
2. McHugh, M.F. Science politics and public policy: a legislator's perspective. *Fedn Proc. Fedn Am. Socs exp. Biol.* **38**, 2567–9 (1979).
3. Gwatkin, D., Wilcox, J., and Wray, J. *Can health and nutrition interventions make a difference?* Overseas Development Council, Monogr. 13, Washington, DC (1980).

Mental handicap

*Gunnar Dybwad**

Mental handicap (or mental retardation) is not a disease but a state, a condition which may be due to a wide variety of causative factors and they may be present singly or in an interactional pattern.

During the last 25 years great strides have been made in identifying and at least partially controlling many of these factors in the biomedical area. However, mental handicap is also the result of environmental, socio-economic-cultural factors and there efforts toward control have been far less successful.

In mental handicap three levels of preventive approaches must be pursued: primary prevention which aims at keeping the condition from occurring at all — if a woman gives up alcohol at the beginning of pregnancy the fetal-alcohol syndrome is eliminated; secondary prevention which through early screening methods facilitates an intervention which prevents mental retardation from developing — an example is testing for PKU (phenylketonuria) and if results are positive, immediate adoption of a restrictive diet; and tertiary prevention which brings to children with mental retardation a regime which holds the disability in bounds, minimizes the damage, and reinforces normal processes — such as will result in Down syndrome from a careful educational programme buttressed by good paediatric attention.

*Professor Emeritus of Human Development, Brandeis University, Massachusetts, USA.

The field of mental handicap was for many decades a stepchild — neither government, nor the institutions of higher learning nor the professional community showed much interest until from the late 1940s to the early 1950s parents of mentally handicapped children rose up in arms and demanded justice for their children. But these parents also recognized the need for scientific enterprise and in the United Kingdom, Canada, the United States, and other countries sponsored and even supported research activities.

It is dangerous to single out any episode but certainly PKU can serve as a good example. It was discovered in Norway by Følling in 1935 and the principle of a corrective dietary treatment was discovered in England in 1953. But it was not until 1959 that the Guthrie test made widescale screening a possibility and the resultant publicity and public education was a real breakthrough — I well remember the astonishment of practising physicians when they first heard about this simply administered test and later heard of its applicability to galactosaemia, maple-sugar urine disease, histidinaemia, and other metabolic disorders. Incidentally, Guthrie's work was supported by the parents' association in the United States.

A few years after the introduction of the Guthrie test came another breakthrough — following two devastating rubella epidemics which resulted in numerous births of severely, multiple-handicapped children, the rubella vaccine was so successful that the disease has been practically eliminated in my State of Massachusetts.

At about the same time another procedure was introduced, amniocentesis, for prenatal chromosomal diagnosis through the drawing of fluid from the uterus of a woman in the second trimester of pregnancy. Thus the presence of Down syndrome and other chromosomal abnormalities could now be ascertained and it was possible for a woman to make an informed decision about termination of her pregnancy.

Another major advance occurred in the late 1960s when a method was developed to detect Tay–Sach's disease in cultured cells obtained via amniocentesis. In this devastating degenerative disease leading to very early and traumatic death a new parental choice, abortion, has thus become available.

Significant progress has also occurred during the past decade in the care of premature infants, many of whom formerly survived as damaged babies with a low level of intellectual functioning. However, while this means that with improved care children who formerly tended in the direction of mental retardation now 'get by', others of still lower birthweight and less vitality often spend extensive periods in intensive perinatal care with doubtful outcome.

This phenomenon of progress bringing with it new setbacks is very much under discussion at this time relative to PKU. While in countries with a well-organized screening and treatment programme the disease has practically disappeared, a new and serious problem has arisen, maternal PKU. When women who have been treated as children for PKU (with a phenylalanine-free diet) and as adults become pregnant, their offspring will almost invariably have severe impairments. The only way to avoid this is a resumption of the diet during pregnancy from the first day, that is to say in practical terms, before conception. It is obvious that this will be a very difficult problem to manage.

Mental handicap is the result of a wide range of causative factors; no less than 100 genetically determined disorders of metabolism alone have been identified. In addition there is the wide array of socio-economic-cultural and environmental factors which may or may not be combined with an organic deficiency. There is still much unexplored territory and this means areas where we cannot yet begin to think about prevention.

In tertiary prevention, i.e. the alleviation of symptoms of disability, results have not been as sharply defined as in secondary prevention, but the results have in the aggregate done much more to change the image of mental handicap. This pertains to successful schooling of children once deemed ineducable, vocational preparation for those once deemed unable to work, community adjustment of those once deemed able to live only in segregated institutions. In addition, early intervention has also resulted in a far higher level of physical activity on the part of persons once considered crib cases and non-ambulatory.

The successes that have been established in tertiary prevention, e.g. in children who have Down syndrome or spina bifida are increasingly bringing forth ethical argumentations as far as prevention is concerned. Since no one can predict at the time the positive results of an amniocentesis are available how far a child with Down syndrome may go in his general development, or more specifically, in his schooling or his level of achievement in daily living, communication and physical skills, or his emotional responsiveness within the family group and without, how much pressure can and should be brought to persuade a mother to have an abortion?

Certainly there is an important qualitative difference here compared with a finding via amniocentesis that the child would be born with Tay–Sach's disease, resulting in an early death following a rapid degenerative process. Some infants and very young children with Down syndrome have in recent years been successfully placed for adoption. And what about the mother whose age suggests that she cannot count on having another pregnancy or in any case another 'normal' pregnancy?

Those advocating termination of any pregnancy involving Down syndrome or who advocate terminating the life of an infant born with Down syndrome claim that such a child has a devastating effect on the life of the family and also refer to the costs of lifetime institutional care. But by now it has been amply demonstrated that such children can live with their family until they have reached the age when they want and can live away from home in an apartment or some other appropriate arrangement.

To be sure, there will be special expenses such as special education, vocational training, etc. but they occur in the cases of many other children who manifest problems of a different nature. We have to consider a much broader question: why this strong negative value judgment about persons with limited intellectual functional level? After all, does not physical health rank as at least of similar importance in the life of the average human being? Are we still facing an aftermath of the eugenic scan of the turn of the century?

In conclusion, what are the implications of all this as far as developing countries are concerned? Obviously, in terms of primary prevention the first needs are clean water, enough food, proper nutrition, basic health care for mother and child — the chance to be born well and to keep well, and in any developing country these are at least governmental priorities.

Next is the need for education, for public information campaigns to fight superstitions and teach health habits, including family planning, and above all, a strengthening of the family and its ties to the community.

These goals are not easily accomplished and indeed sound like a fantasy to a country steeped in poverty. But in many developing countries regions of abject poverty adjoin cities with a vastly higher level of facilities and standard of living. There, services can be demonstrated and from there some modest outposts can be set up in the as yet unserved areas.

In the industrialized countries an undue emphasis on technical services and professional standards led to a devaluation of the family's basic nurturing role. More recently the key importance of this role of the family in dealing with childhood disability has been rediscovered. Developing countries have a chance to build any programme dealing with disability around the family, using its strength to extend care to the disabled child and providing for that child, as much as possible, opportunities for normal growth and development within its own societal patterns.

It is from families that there can come and has come the initiative to start programmes under their own power when no one else is willing to provide service. Parents of handicapped children have shown a rather similar pattern of action throughout the world; first the worry about their own child; initiation of a simple home-based programme (usually schooling); inclusion of other children; reaching out for professional help; a campaign for public programmes buttressed by a formally organized association of parents and friends; and finally insistence on preventive services.

Voluntary associations such as are organized in the Council of World Organizations Interested in the Handicapped (CWOIH), the United Nations through UNICEF, and the several specialized agencies such as WHO, UNESCO, ILO, and FAO, all have developed effective means to support such initiatives in developing countries until the government has the strength to provide the needed help and guidance.

3

The principal causes – acute conditions

Blindness
*Sir John Wilson**
Prevalence

According to the definition used, there are estimated to be from 28 to 42 million blind people in the world (World Health Organization 1980). Because blindness is linked to aging and to population growth it has been estimated that, if present trends continue, the number of the blind in the industrialized countries will double over the next 50 years, and will multiply fivefold in the developing countries (US National Eye Institute 1978).

Objective

The principal aim of the international prevention of blindness programme — which is a priority of WHO's global technical co-operation programme — is, over the next 20 years, to control four major causes of blindness which together account for two-thirds of the avoidable blindness in developing countries. Intergovernmental programmes are now in operation in five of the six World Health Regions. National programmes are operating or have been formulated in 45 developing countries. Linked to the International Agency for the Prevention of Blindness (IAPB), which was established to co-ordinate non-governmental action, National Committees now exist in 56 countries. WHO has identified 11 'Collaborating Centres' for research and staff training.

Priorities

Before the World Health Assembly adopted the global programme, an impressive consensus of expert opinion had concluded that an appropriate technology exists to control, at an acceptable level of cost-effectiveness, the four major causes of blindness, which subsequently became priorities of the programme. These causes are: trachoma and the communicable eye diseases; onchocerciasis (river blindness); xerophthalmia (blinding malnutrition); and cataract. The aim is not eradication, which would be impracticable. It is 'to eliminate the over-burden of avoidable blindness' in those communities identified as having excessive prevalence, i.e. in over 1 per cent of the population.

*President, International Agency for the Prevention of Blindness, Director, Royal Commonwealth Society for the Blind.

This involves supplementation of primary health measures against trachoma and xerophthalmia, and specific interventions against onchocerciasis and cataract.

Trachoma

The expectation is that, in any area of major prevalence, a systematic control programme combining individual therapy with environmental health measures should be capable, in five years, of reducing the blindness rate to the point where control can be continued as a regular feature of community health. Such programmes have already been proved effective in parts of the Middle East, Southern and South-East Asia, Soviet Central Asia, Australia, and areas of Africa and Latin America. Earlier hopes of a vaccine have been frustrated but the potential for control continues to improve with the development of better systems of drug delivery, clearer identification of at-risk groups, and of environmental factors.

Onchocerciasis

The West African Onchocerciasis Control Programme, now in its second five-year phase, operates in seven West African countries, with support from four UN Agencies and many governments. By vector control and related measures, transmission of the disease has now been interrupted over 80 per cent of the control area of 720 000 km^2. Recent evaluation found no trace of the disease in any child born since transmission was interrupted, by comparison with some 20 per cent of children affected in non-control areas. Diminution was observed in severity of the disease amongst affected adults, but no satisfactory individual therapy has yet been demonstrated for that substantial fraction of the population which acquired a blinding infection before control was instituted. Though many problems remain — notably preventing reinvasion and achieving cheaper means of permanent control — optimism seems justified that within 20 years this cause of mass blindness can be effectively controlled in its major endemic areas.

Xerophthalmia

Against this disease — associated with vitamin A deficiency and now recognized as the largest cause of child blindness in developing countries — three interventions have been demonstrated. These are: food fortification (Central America); distribution of vitamin concentrate (Southern and South-East Asia); nutritional rehabilitation, resulting in systematic use of dark-green leafy vegetables (DGLV) in children's diet (India and Indonesia). Such measures, with identification of children at risk, can be delivered inexpensively through primary health care, with the expectation that wherever the necessary behavioural change can be sustained, xerophthalmia can be eliminated as a major cause of child blindness.

Cataract

Throughout much of Asia and Africa, 40 per cent of the blindness is attributed to curable cataract. Techniques of mass, low-cost surgery (in Eye-camps; mobile units; regular or improvised hospitals) are well established. In areas of major prevalence 'crash programmes' of sight restoration aim at clearing the backlog to the point where regular hospital services can handle the annual load of new cases. Limiting factors include: the size of this backlog (some five million cases in India); shortage of surgeons, particularly in rural areas; reluctance to assign a sufficiently responsible role to auxiliaries. Research possibilities might include: search for a safe and economic technique of mass out-patient treatment; systems study to maximize resources; better understanding of the chemistry of cataract formation, aimed at postponing its onset.

Other priorities

The global programme focuses attention on these four diseases because their control, through an already deliverable technology, could decisively reduce blindness throughout the world. Other causes are receiving attention regionally and nationally.

GLAUCOMA

Drug therapy, recently significantly improved by the introduction of beta blockers, is extensively controlling but not curing glaucoma in industrialized countries. This therapy is at present too expensive and impractical for mass delivery in developing countries, though surgical intervention for acute glaucoma is increasingly practised even in Eye-camps.

TRAUMA

Trauma is a major cause of blindness in industrialized countries, where prevention and treatment is increasingly effective. In developing countries much could be done to prevent agricultural accident and to establish eye banks.

RUBELLA

Control should be an essential feature of immunization programmes.

LEPROSY

Improved surgery and stronger motivation could prevent much blindness.

AGE-LINKED DISEASES

Considerable research resources are devoted, though with varying estimates of an early outcome, to diabetic retinopathy and macular degeneration — which contribute largely to blindness in aging populations — and to hereditary and congenital causes of blindness.

Cost-effectiveness

WHO has described the prevention of blindness as one of the most cost-effective areas in international health. A joint Pan-American Health Organization/World Bank meeting (Washington, July 1980) estimated at US $ 350–560 M annually the cost of adding an effective eye-care component to primary health services throughout the developing world.

Deafness

Yash Pal Kapur*

Deafness and profound impairment of hearing affect an individual's ability to communicate with his fellow beings. Impaired hearing disrupts an individual's integration into the human community. This is a hidden disability and, unlike other handicaps, much less is known of its aetiology and, in many countries, its prevalence.

Prevalence of impaired hearing

Significant degrees of hearing loss affect about 10 per cent of the population. This value applies to Australia, Europe, and North America. Much higher values apply to countries in Africa, the Middle East, and South-East Asia.

A survey conducted on behalf of the Social Commission of the World Federation of the Deaf in 1975 gave the prevalences for hearing impairment and deafness in 28 countries. Estimates of hearing impairment range from 265 (0.265 per cent) to 6603 (6.6 per cent) per 100 000 of the general population in these countries. Estimates of the deaf population in 20 countries range from 13 (0.013 per cent) to 873 (0.873 per cent) per 100 000 general population (see Table 3).[1] Appendix 3.1 shows the classification of the hearing impaired into the deaf and the hard of hearing.[2]

*Professor of Otolaryngology, Michigan State University, USA.

Table 3.1 Prevalence rates for hearing impairment and deafness per 100 000 general population: estimates from 28 nations, 1960–75[a]

Nation	Year[b]	Hearing-impaired	Deaf
United States	1971	6603[c]	873[c]
Australia	1974	4155	——
Egypt	(1975)	3369	——
Ethiopia	1975	1917	——
Philippines	(1974)	1090	——
Brazil	1975	983	——
Panama	1974	——	554
Ireland	1974	337	——
Great Britain[d]	(1975)	306	——
Tunisia	1971	265	——
South Africa	(1975)	——	261
New Zealand	1974	——	229[e]
Japan	1970	——	225
Liberia	1971	——	191
Turkey	1965	——	176[f]
Taiwan	(1974)	——	152
Finland	(1974)	——	140
Kenya	(1974)	——	137
Uganda	1969	——	99
Sri Lanka	1971	571[g]	97[e]
Portugal	1960	——	82
West Germany	(1975)	——	82
Denmark	(1974)	——	81
Ghana	1974	——	60
Belgium	1974	——	51
Mexico	1974	——	46
Luxembourg	(1975)	——	27
Thailand	1974	——	13[e]

[a]Estimates for general population are for ages ranging from infancy to over 85, unless otherwise notes. Source: United Nations (1974).

[b]Dates in parentheses refer to gross estimates of hearing-impaired or deaf populations made by respondent. Other dates refer to published data or official source or basis of population estimate.

[c]Persons ranging from infancy to age 65.

[d]England and Wales only.

[e]School-age deaf children.

[f]Deaf children 6–12 years of age.

[g]School-age hearing-impaired children.

Prepared by Martin McCavitt.

The areas of major concern

CONGENITAL DEAFNESS

From 10 to 26 per cent of cases of hearing loss in children are known to be congenital, that is, present at birth or by early childhood.[3] Congenital deafness may be acquired or genetically caused.

Acquired congenital deafness

Exogenous factors which result in hearing loss are maternal viral infections, ototoxic agents, kernicterus, and perinatal disorders including
prematurity, birth trauma, or hypoxia.

Ototoxic agents that cross the placenta may affect development of
the inner ear. The damage is usually sustained during the first trimester.
The aminoglycosidic group of antibiotics are prime offenders. Salicylates, chloroquine, and quinine taken during pregnancy are also
ototoxic.

Rubella virus infection leads the list of maternal viral infections
which result in severe hearing loss. The rubella virus involves other
organs particularly the hearing and the eyes. Other complications that
result are heart defects and mental retardation. Congenital rubella has
been estimated to cause between 5 and 20 per cent of all cases of congenital deafness in epidemic years.[4]

The rubella virus is known to be a virulent destructive agent to fetal
organs when it affects the mother especially during the first trimester
of pregnancy. It is generally accepted that 25–30 per cent of the
children of mothers who had rubella in the first few weeks of pregnancy
have a hearing loss. These figures are based on recognition of overt
rubella in the mother. However, it is now known that subclinical infection can occur resulting in maternal rubella without a rash and may
be responsible for unexplained congenital deafness.[4]

Congenital rubella is a preventable disease. Immunity conferred
either by vaccination or by natural infection prevents transplacental
infection of the developing fetus. The United States' experience is en-
lightening. Since the rubella vaccine was licensed in 1969 and the
rubella vaccination began, there has been a dramatic decrease in cases
of rubella deafness. The efforts of rubella vaccination have focused
primarily on immunization of pre-school and young school-age children
to decrease overall rubella cases and consequently the chances of ex-
posing susceptible postpubertal females. Vaccination of postpubertal
females was given lower priority. Implementation of this strategy has
resulted in an interruption of the characteristic rubella epidemics at
six- to nine-year intervals, a marked reduction in the incidence of
rubella, and prevention of epidemics of congenital rubella syndrome.
In young school-age children, the rubella incidence decreased 89 per
cent between 1966–1968 and 1975–1977. These figures testify to the
effectiveness of rubella vaccination programmes.[5] The congenital
rubella syndrome with its sequelae of deafness, cataracts, mental retardation, and heart murmurs imposes a great demand on habilitative,
educational, and social services which very few countries can afford to
bear. Prevention is possible and immunization has proved to be effective.

Other viral infections causing deafness are the viruses of mumps,
measles, varicella-zoster, and the cytomegalic virus. Immunization

programmes can prevent mumps, measles, and varicella-zoster infections.

The ultimate goal of those who work in the field of deafness is prevention. If in 50 per cent of profound childhood deafness the aetiologic factor is due to exogenous causes, then these cases can be prevented.[6] Knowledge of drugs which are and can be potentially ototoxic needs to be publicized among the health professions.

Hyperbilirubinaemia and kernicterus can be prevented by light therapy and exchange transfusions.

Improved management of perinatal problems will lessen the hearing loss resulting from birth complications.

Genetic/congenital deafness

Fifty-two per cent of congenital deafness is genetically caused.[3] Estimates of the incidence of congenital deafness in Great Britain, Japan, Germany, and the United States are one in 2000 to one in 6000 live births.[7] The majority of genetic deafness is transmitted through recessive genes. Recessive deafness is estimated to account for 75–88 per cent of genetic deafness. Dominant deafness comprises 12–25 per cent of genetic deafness. Hereditary deafness may exist alone or be associated with other defects. In 1969, Konigsmark listed 70 types of hereditary deafness in man.[8] Many more syndromes have been added to this list since then. The custom of consanguineous marriages present in large parts of the world contributes to the particularly high prevalence of hereditary deafness in the Moslem world in the Middle East, parts of Africa, and South-East Asia.[9]

Profound hereditary deafness is a more serious handicap than most deafness as therapy is not possible apart from management by educational measures. In this situation, special stress has to be laid on prevention. Prevention of genetically-determined deafness comes within the province of genetic evaluation and counselling. The accuracy of the counselling is dependent on the accuracy of the diagnosis. About two-thirds of genetic deafness is due to the autosomal recessive mechanism. In these cases, counselling can only be retrospective, i.e. after the birth of an affected child.

Preventive steps include education of the health-care providers in this area and of provision of centres for genetic evaluation and counselling.

Genetic counselling is given to parents who have had one abnormal child and who are interested in knowing the potential for having additional children with the same defect. Genetic counselling may also be offered to siblings of deaf individuals and to the affected persons themselves as they approach marriageable age and possibly parenthood. There is good evidence that parents who seek genetic advice will usually make appropriate and expected decisions about future children.

The field of genetics is old but it is still not well understood by health professionals nor is it well stressed in the curriculum of medical schools. The advances in the science of genetics and genetic screening programmes have brought an explosive expansion of prenatal screening and genetic counselling services. Understanding by the health professionals of the basis for genetic transmission of disease and ability to construct a family pedigree is helpful in understanding the modes of inheritance. These types of knowledge will increase referrals both for prenatal screening and genetic counselling leading to decreasing the prevalence and impact of genetic diseases on the hearing.

No discussion of genetic counselling would be complete without the mention of prenatal diagnosis. Prenatal genetic evaluation is undergoing rapid technological advances and an ever-increasing number of genetic diseases are being diagnosed before birth. Prenatal monitoring is accomplished at the present time through three techniques: amniocentesis, ultrasound, and fetoscopy.

Risk registration for children suspected to have hereditary deafness has made significant progress in many countries.

Otitis media

Otitis media is the most common cause of hearing loss in children. Infants and young children are at the highest risk for acquisition of otitis media with a peak prevalence rate between 6 and 36 months.[10] It results from upper respiratory and systemic infections. The hearing loss is usually conductive and reversible by medical and surgical treatment.

Teele *et al.* found that by age three years more than two-thirds of the children in their study had at least one episode of otitis media.[11] The overall prevalence rate in children has been estimated to be between 15 and 20 per cent.[12]

Surveys have shown several risk groups. Socioeconomic factors are predisposing conditions which influence morbidity rate and severity of otitis media. Poor sanitary facilities, lack of adequate nutrition, and severe overcrowding are a high-risk environment for acute and chronic infection of all types. Racial factors have also been described. The Eskimos and American Indians demonstrate disproportionately high rates of middle-ear disease. It was less common in Blacks.[13] Other high risk factors are children with cleft palate and other craniofacial abnormalities and children with Down syndrome.

It is estimated that 5 per cent of the children in the public school systems in the United States have impaired hearing with otitis media being the major factor. Studies from other countries show the following figures for hearing loss in children:

United Kingdom	3.5–6.6 per cent (Watson; Dunn)[14] [15]
Denmark	3.5 per cent (Fabritus)[16]
India	17 per cent (Kapur)[17]
Kenya	15 per cent (Clifford)[18]
Uganda	15 per cent (Roland)[19]

Left untreated, otitis media can have extremely serious behavioural as well as physical consequences. The two areas of development most affected are language development and educational achievement. It has been suggested that even a mild hearing loss (15 dB or less) in a young child over a prolonged period may adversely affect language and educational development. The children in the school systems are the future of every nation and this handicap is detrimental to their future performance. Early identification and early intervention can result in reversal of the hearing loss. Preliminary results of the efficacy of pneumococcal vaccination against recurrent otitis media in Finland and the United States have been encouraging. A study of 781 children ages 3–83 months with acute otitis media were immunized with either pneumococcal or *Haemophilus influenzae* vaccine. Follow-up of one to 17 months, average of 13 months, showed a significant decrease in the recurrence of otitis media in any of these children.[20] Preventive measures against the common childhood infectious diseases, such as immunization measures, appear to be the most logical and practical way of preventing otitis media.

If acute otitis media is inadequately treated or neglected, chronic inflammation results in a chronic otitis media which can be corrected only with surgical treatment. Left untreated, otitis media can have extremely serious consequences including labyrinthitis and meningitis which cause severe deafness.

Meningitis and hearing loss

Bacterial meningitis is a leading cause of acquired sensorineural hearing loss. Several studies of school-age children have shown that bacterial meningitis is the cause of 8–24 per cent of all cases with sudden deafness. Among the cases of deafness attributed to acquired postnatal causes, the incidence of bacterial meningitis is even higher. The incidence of hearing impairment as a sequela of bacterial meningitis ranges from 5 to 35 per cent of survivors.[21]

In West Africa in a 'meningitis belt' that extends in the northern part, meningococcal, pneumonococcal, and viral meningitis were considered to be responsible for 18 per cent of the cases of postnatal acquired deafness in Nigeria.[22] The incidence of meningococcal meningitis has increased worldwide over the last 10 years and continues to be responsible for a large number of cases of deafness.

Most meningitis hearing losses are bilateral and most of them produce a severe irreversible loss. Many patients suffer temporary or partial hearing loss that may go undected unless there is a hearing evaluation. The three major organisms responsible for meningitis in children are *Haemophilus influenzae*, *Neisseria meningitidis*, and *Diplococcus pneumoniae*.

Early identification and treatment of otitis media by medical and surgical means and early diagnosis and initiation of medical therapy in meningitis are important to decrease the incidence of postmeningitic hearing loss and neurological sequelae.

Hearing loss from noise

Industrialization and modern technology have created many environmental pollutants of which noise is an immediate and identifiable example. Long-term exposure to noise results in a sensorineural hearing loss. In the United States, a conservative estimate places the number of people in industry with a noise-induced hearing loss at five or six million.[23] The maximum allowable noise exposure to industrial noise is eight hours per day at 90 dB.[24] Recent studies demonstrate that excessive noise exposure cannot only result in a hearing loss but can also significantly affect the work performance of the worker. Other non-auditory effects of noise include a greater incidence of hypertension and psychosocial abnormalities. Animal experiments show that excessive noise exposure can be teratogenic resulting in birth abnormalities.[25] Many Western countries have adopted laws which list occupational deafness as a prescribed industrial disease and have laid down standards for noise attention and hearing conservation.

The problem, however, is worldwide. The worker in many parts of the world has no protection from noise. There is a need for the education of industry, governmental agencies and the public on the effect of noise on the hearing. This type of hearing loss can be prevented by noise reduction and ear protection. The cost of hearing conservation is far less than the cost of hearing loss in terms of human suffering and the cost of treatment and management of hearing loss. Proper education of both management and labour can result in successful hearing conservation programmes. The method of choice is reduction of the noise level at the source. Should this not be feasible, the hearing should be protected from the noise. Experience has shown that hearing protection programmes can prevent hearing loss in the majority of individuals exposed to excessive noise.

Ototoxic drugs

The list of agents known to have ototoxic effect is already prodigious and undoubtedly will continue to grow with more pharmaceutical advances. A publication compiled by the Johns Hopkins' Medical Institution listed nearly 100 ototoxic products that had been documented in the medical literature between 1966 and 1971. Since that time, the list has doubled.[26]

The list of ototoxic agents include many antibiotics, diuretics, analgesics, antipyretics, antineoplastic chemicals, and topically-applied agents. Quinine, salicylates, and streptomycin were the earliest known ototoxic agents. Many of these still continue to be used.

Many of these agents cause delayed hearing loss. The hearing loss may be detected weeks or months after cessation of the drug therapy. The young and the very old seem to be a greater risk. The pregnant woman is a special risk.

With the proliferation of pharmaceutical agents, the medical practitioner and health-care providers need to be provided with updated information on the adverse and toxic effects of these agents.

Nutritional deficiencies and hearing loss

Kwashiorkor, a protein/calorie malnutrition, exists in many countries of Africa and Asia. Histopathological studies of the temporal bones taken from kwashiorkor patients show severe damage to the middle and the inner ears and to the vestibulocochlear nerve.[27] The Nigerian oto-neuro-ophthalmological syndrome is due to chronic cyanide poisoning. This arises from the consumption of cassava.[28] Nutritional inadequacies are widespread throughout the world. Their impact on the auditory system needs more study and research.

Other priorities

EARLY IDENTIFICATION

Early identification of hearing impairment is the key to medical and surgical treatment, habilitation, and education. The ideal age for detection is within the first few months of birth. In Denmark, where suspected infants are followed closely, the average age at which this hearing handicap is discovered is age two years. In developing countries, detection occurs at an average age of six years. This is very late and has serious social, educational, and habilitative implications. Children suspected of having a hearing loss should be screened at ages seven to nine months. The examination should include an otoscopic examination to detect middle-ear disease. Hearing evaluation should be focused on children with the following risk criteria:

1. Family history of hearing loss.
2. Rubella or other viral infections in the first half of pregnancy.
3. Anatomical malformation involving the head and neck.
4. Birth weight less than 1500 g.
5. Serum bilirubin exceeding 20 mg per 100 ml serum.
6. Bacterial meningitis.
7. Severe asphyxia.[29]

Public Health Departments should provide training for paediatric nurse practitioners, public health nurses, and other designated health personnel in performing behavioural screening tests for detecting hearing loss in children.

Summary

The major problem in the area of hearing loss is the lack of information, especially on measures to prevent the handicap and the importance of early diagnosis and remediation. An international centre for dissemination of such information could serve a long-felt need.

Appendix 3.1

Richard Silverman

1. *The Deaf*: Those in whom the sense of hearing is non-functional for the ordinary purposes of life. This general group is made up of two distinct classes based entirely on the time of the loss of hearing:
(a) The congenitally deaf: those who are born deaf.
(b) The adventitiously deaf: those who were born with normal hearing but in whom the sense of hearing becomes non-functional later through illness or accident.

2. *The hard-of hearing*. Those in whom the sense of hearing, although defective, is functional with or without a hearing aid.

References

1. McCavitt, M. International survey of the prevalence of deafness and services to deaf people. *Proceedings of the Congress of the World Federation of the Deaf*, p. 469. Washington, DC (1975).
2. Silverman, S.R. and Lane, H.S. In *Hearing and deafness*, 3rd edn (ed. H. Davis and S.R. Silverman) p. 384. Holt, Rinehart and Winston, New York (1970).
3. Steele, M.W. In *Pediat. Clins N. Am.* **28**, 973 (1981).
4. Karmody, C. Subclinical maternal rubella and congenital deafness. *New Engl. J. Med.* **278**, 809 (1968).
5. Orenstein, W.A. and Greaves, W.L. Congenital rubella syndrome: a continuing problem. *J. Am. med. Ass.* **247**, 1174 (1982).
6. Fraser, G.R. Profound childhood deafness. *J. med. Genet.* **1**, 118 (1964).
7. Fraser, G.R. In *The causes of profound deafness in Childhood*, p. 281. Johns Hopkins University Press, Baltimore (1976).
8. Konigsmark, B.W. Hereditary deafness in man. Medical progress, Massachusetts Medical Society. *New Engl. J. Med.* **281**, 713 (1969).
9. Kapur, Y.P. Study of aeteology and pattern of deafness in a school for the deaf in Madras. *Proceedings of the Congress of the World Federation of the Deaf*, 1967, Warsaw, Poland (1968).
10. Kessner, D., Snow, C.K., and Singer, T. *Assessment of medical care for children: contrasts in health care status*, Vol 3. Institute of Medicine, National Academy of Sciences, Washington, DC (1974).
11. Teele, D.W., Kelin, J.O., and Rosner, B.A. Epidemiology of otitis media in children. *Ann. Otol. Rhinol. Laryngol.* **89**, Suppl. 68, 5 (1980).
12. Avery, A.D. *et al.* Quality of medical care assessment using outcome measures: eight disease-specific applications. Prepared for the Health Resources Administration, Department of Health, Education and Welfare by the Rand Corporation: Santa Monica (1976).

13. Bluestone, C.D. In *Nelson's textbook of pediatrics*, 11th edn (ed. V.C. Vaughan, R.J. McKay, and Behrman) p. 1180. Saunders, Philadelphia (1979).
14. Watson, T.J. Personal communication (1964).
15. Dunn, M.W. Audiometric survey unit — the Glasgow School Health Services. *Hlth Bull.* XXI, No. 2 (1963).
16. Fabritius, H.F. Nine years examination of the hearing in school children in North Trondelag. *Acta oto-lar.* **188**, Suppl., 351 (1964).
17. Kapur, Y.P. A study of hearing loss in school children in India. *J. speech hear. Dis.* **30**, 225 (1965).
18. Clifford, P. Causes of Deafness in Africa. *Report on the First Seminar on Deafness to be held in Africa*, Nairobi, Kenya. The Commonwealth Society for the Deaf, London (1968).
19. Roland, P.E. Otological problems in Uganda. *J. Laryngol.* **74**, 678 (1960).
20. Karma, P. *et al.* Efficacy of pneumococcal vaccination against recurrent otitis media: preliminary results of a field trial in Finland. *Ann. Otol. Rhinol. Laryngol.* **89**, Suppl. 68, 357 (1980).
21. Nadol, J.B. Hearing loss as a sequela of meningitis. *Laryngoscope* **88**, 739 (1978).
22. Holborow, C., Martinson, F., and Anger, N. In *A study of deafness in West Africa*. Occasional Paper by the Commonwealth Society for the Deaf, London (1981).
23. Heffler, A.J. In *Otolar. Clins N. Am. Audiol.* **11**, 723 (1978).
24. Schiff, M. In *Sensorineural hearing loss, vertigo and tinnitus* (ed. M.M. Paparella and Meyerhoff) Ear Clinics International, Vol. 1, p. 59. Williams & Wilkins, Baltimore (1981).
25. Geber, W.F. Developmental effects of chronic maternal audiovisual stress on the rat fetus. *J. Embryol. exp. Morphol.* **16**, 1 (1966).
26. Quick, C.A. In *Otolaryngology*, Vol. II. *The ear* (ed. M.M. Paparella and Shumrick) p. 1804. Saunders, Philadelphia (1980).
27. Kapur, Y.P., Michaels, L., and El-Sheikh, M. Histopathological changes in the temporal bone in kwashiorkor. *Clin. Otolar.* **5**, 171 (1980).
28. Hinchcliffe, R. In *Scientific foundations of otolaryngology* (ed. R. Hinchcliffe and D.F.N. Harrison) p. 133. Heinemann, London (1976).
29. Downs, M.P. In *Otolaryngology* (ed. C.M. English) Vol. 1, p. 1. Harper & Rowe, Hagerstown (1980).

Injury

Leo A. Kaprio*

Size of the problem

Consideration of injury prevention necessitates the study of its causes. Accidents, poisoning, and violence represent new and insufficiently recognized modern health hazards in the world, with a real pandemic character and manifestation.

Evaluation of the magnitude of the problem would be possible if adequate statistical information were available. At the international level, available and readily accessible information is limited to data on

*Director, Regional Office for Europe, World Health Organization.

mortality, on injury and death caused by road traffic accidents, and on fatal industrial accidents.

MORTALITY

In the absence of appropriate morbidity data, mortality statistics are used regularly for indirect evaluation of the health status of populations and their health problems. The statistical data for 58 countries show a total of 705 687 deaths from accidents, poisoning, and violence for a population of 1 104 835 000, i.e. a death rate of 64 per 100 000.

Accidents, poisoning, and violence generally rank third as a cause of death in developed countries, and in many developing countries they are assuming increasing importance and moving up in the ranking. The mortality rate per 100 000 population in males varies widely from 50.9 to 148.8 in developed countries, and from 33.0 to 135.8 in developing countries; for females the respective ranges are 23.6 to 82.7 and 16.0 to 45.1.

Death by accidents, poisoning, and violence as a percentage of all deaths varies widely for males from 4.1 to 12.2 per cent in developed countries, and from 5.2 to 20.1 per cent in developing countries; for females the rates are 3.1 to 7.7 per cent and 2.7 to 9.1 per cent, respectively.

In the developed countries, in the age group 1–44 years (which has a relatively low general mortality), accidents occupy first place. A recent study showed that, for 50 countries studied, the average accident mortality in those aged 1–4 years was 32.7 per 100 000 boys and 22.8 per 100 000 girls, dropping to 23.1 for boys and to 10.9 for girls in the age group 5–14 years.

Industrialized countries that have succeeded in considerably reducing not only the general but also the accident mortality rates corroborate the thesis that accidents can be reduced in all countries. Apart from children it is well known that accidents in the elderly cause a high percentage of mortality. In Europe accidents to elderly pedestrians are one of the greatest causes of mortality and morbidity.

MORBIDITY

Morbidity statistics in this field are poorly developed, but it is estimated that for one death on the road there are about 15 severe and 30 slight injuries. In England and Wales in 1975, for every accidental death about 26 people were hospitalized or discharged because of accidents.

TRENDS

In the developed countries the number of accidents and consequently deaths and injuries have declined, but because of the improving level of health in those countries, the relative importance of accidents as a cause of death and injury is increasing, especially in the young and the elderly. In developing countries, with increasing industrialization, the numbers of accidents and, consequently, deaths and injuries are increasing, and unless preventive measures are taken as the health situation improves the contribution to death and morbidity from accidents will increase in relative terms.

THE COST TO SOCIETY

The economic cost of road accidents is estimated at 1 per cent of GNP on an average in developed countries. In England, the cost to the health services from road accidents alone is estimated at some £75 million a year, the cost from home accidents is about £87 million and the total cost from all accidents is assessed at well over £400 million.

From the health services standpoint, many of the beds that could be used for urgent cases are occupied by victims of accidents, especially road traffic accidents. To this must be added loss of working time, provision of social security and disability benefits, and provision of special education and treatment.

A programme for prevention

WHO's concern with road accidents world-wide was expressed for the first time during the Nineteenth World Health Assembly in 1966, when the Member States passed a resolution requesting WHO to play a more active role in the prevention of accidents. Since then the European and American Regions in particular have been actively engaged in developing specific regional programmes, and the other Regions are now also paying increasing attention to this problem. In 1976 the Regional Office for Europe was given responsibility for developing the WHO global programme for the prevention of road traffic accidents, in addition to its own growing regional programme.

PROGRAMME DEVELOPMENT

Most of the improvement in the situation has been undertaken by authorities outside the health sector, and often without support from the health sector. Health authorities have tended to concentrate on the development of accident and emergency services for the rescue and treatment of accident victims. In developing prevention programmes we must determine how health authorities can support those who have the main responsibilities in this field. To do this they need to consider how they can organize themselves and include the prevention of accidents in their own preventive health programmes.

POTENTIAL FOR PREVENTION

Accidents are preventable. This has been demonstrated by the successful efforts in many countries in Europe. Any programme should comprise primary (prevention of the accident), secondary (alleviating or mini-mizing any trauma that occurs), and tertiary (provision of long-term care) elements.

STRUCTURED PROGRAMMES

There are many programmes on accident prevention. Indeed, there are so many agencies involved world-wide that one of the main problems is intersectoral co-ordination. Most of these programmes do not have health-sector involvement; WHO has therefore developed a programme in its European Region which is now about to be used as a basis for the global programme (beginning with a major conference in Mexico for accidents in developing countries).

The European programme

As with all WHO programmes the strategy and objectives were developed through the use of expert committees and working groups, and more importantly through support for country programmes. The following areas of activity could form a basis for both strategy and action by the health sector.

INTERSECTORAL CO-ORDINATION

Accident prevention programmes are fragmented and the responsibilities of the public-health authorities are ill-defined. Efforts are needed to promote a multidisciplinary and intersectoral approach. The Regional Office has attempted to co-ordinate activity by holding meetings with all the NGOs and the IGOs concerned. To improve this co-ordination further proposals will be made to establish a European task-force of health and other agencies (IGOs and NGOs) involved in accident pre-vention, which will meet biennially to review activities and prepare a joint programme. The Office will encourage similar intersectoral co-ordination at country level.

INFORMATION AND STATISTICS

The planning, implementation, and evaluation of accident prevention and care programmes is hampered by inadequate and inappropriate statistical indicators and deficient mechanisms for the collection and analysis of data. The programme so far has studied methods of data collection and has funded and co-ordinated a study on the collection of data within health services. Recognizing that there are a large number of different indices in use relating to severity of injury, a meeting will be held during 1982 involving health service personnel responsible for primary care, accident, and emergency traumatology and rehabilitation; medical statisticians; epidemiologists; and experts from those agencies that have the main responsibility in the fields of transport, home, sports, and industrial accidents. One of the main objectives will be to develop common indicators and a simple methodology that can be used in developing countries.

HEALTH FACTORS AND HUMAN BEHAVIOUR

In spite of the excellent epidemiological work that has been done so far there is an important gap in knowledge relating to the behavioural (human) aspects of accident causation. In this field the programme has considered the role of health factors and studied the role of alcohol and drugs and psychosocial factors in accident causation. Health services and their personnel have a great deal of knowledge and experience in the behavioural field, and it is therefore essential that they become heavily involved with their colleagues in other sectors. It is hoped that by 1984 an epidemiological review of accidents will be published which will provide a basis for the development of programmes. Studies in countries will be supported by the Regional Office, and it is hoped to publish the results of the first phase of the WHO Co-ordinated Research Projects on Alcohol, Drugs, and Driving in 1985.

THE ENVIRONMENT

In this area help has been provided by the health sector. Indices of severity of injury are urgently needed to inform engineers and environmental scientists and consumer protection agencies, not only to indicate priorities but also to evaluate the effectiveness of countermeasures in injury protection. There are several examples of environmental modification that have reduced the effects of accidents, for example fire-resistant night clothing and child-proof containers. Restraint systems in cars and crash helmets have reduced severe disability due to head and spinal-cord injuries.

The Regional Office has supported a study in biomechanics and intends to support activities relating to the environmental aspects of road use, home accidents, sports injuries, and industry.

'AT RISK' GROUPS

The Office has financed a study on the prevention of traffic accidents in childhood, and during 1982 it is intended to hold a technical discussion on accidents in the elderly. WHO will be proposing that countries establish national co-ordinating committees for accident prevention in children and the elderly. There is a close working relationship between the Regional Office and the ICC, and individual projects within countries aimed at children are being developed. Most countries have well-developed child health services and the inclusion of prevention programmes for accidents to children should not present difficulties, given the necessary professional and political will.

EDUCATION IN ACCIDENT SAFETY

Although safety education is the responsibility of agencies outside the health sector, most health authorities have or are developing health education programmes. The Regional Office financed a study on education in traffic safety in Europe and a technical group discussed this problem in Essen in 1979. Further work needs to be done and the Office will be proposing that, starting in 1984, biennial meetings be held between health and safety educationalists to develop comprehensive programmes and also to establish training programmes for the personnel involved. The development of a standard manual for training programmes for professional staff engaged in accident prevention would be a major step forward in co-ordination.

RESEARCH

There is a special need for promotion and better co-ordination of research within the public health sector as well as closer co-operation between public health and other sectoral research. The programme so far has stimulated several studies and WHO has designated the Transport and Road Research Laboratory in the United Kingdom as a collaborating centre. It is proposed to establish more contractual service agreements with national research institutes over the whole field of accidents, and a review of national programmes of research will be carried out to identify priority areas for international co-operation.

PLANNING AND ORGANIZATION OF EMERGENCY SERVICES

Health-care services for the treatment of accident victims, particularly in emergency situations, are inadequate and often involve the use of inappropriate medical technologies. It is important to promote the planning, implementation, monitoring, and evaluation of *appropriate* health-care services for the treatment of the injured. A balance has to be struck between the introduction of simple self-help measures in developing countries and the correct application of appropriate technology. Following a technical group discussion in Toulouse, a study on the use of appropriate methodology and technology has begun in Morocco which will be of assistance to developing countries. For the future, biennial meetings with health departments will be organized in an endeavour to rationalize country planning for emergencies and especially to involve communities and encourage self-help in support of high technologies. Guidelines for curriculum development in the treatment of accident cases for use by health personnel, especially those involved in primary health care, should be produced, and advances in technology will be reviewed and assessed and the results published.

LEGISLATION

Health authorities should support in all ways those proposals for legislation which are considered essential by the responsible national agencies. The health service input is admirably demonstrated by the support given by the health professions in the various countries to seat-belt legislation. In the fields of home and industrial safety, epidemiological input from the health sector has helped in the development of protection standards.

LONG-TERM EFFECTS OF INJURY-CAUSING ACCIDENTS

Attempts have been made to estimate the burden on society arising out of injuries from accidents. There is a need to establish comprehensive and agreed indices of injury severity which can be used by all agencies, not only within the health sector but from the transport, home safety, and industrial sectors. This is important for the future planning of rehabilitation services.

Occupational accidents

*George Foggon**

The protection of workers against accident and disease in their working environment is a serious and growing problem. It is estimated that about 160 000 industrial accidents occur daily, many resulting in permanent disability. Many of those affected are, moreover, women and young persons — even children. Increasing industrialization in the developing countries of the world exposes the workforce in many countries to new risks and dangers for which there is little effective protection, either by law or by administrative oversight. Indeed there are indications that certain processes which are repudiated by workers in industrialized countries (for example, working with asbestos) may be transferred to countries where regulations are less onerous. In 1980 a manufacturer in the Federal Republic of Germany appealed to the High Court about unfair competition from asbestos products imported from a newly industrializing country, whose requirements for the protection of the workers making them was allegedly negligible. The complainant claimed that the stringent precautions adopted to protect its own workers added about 25 per cent to costs.

In the newly industrializing countries, new technologies are continually being adopted and there is usually rapid and successful reaction in seeking solutions to new production problems; but this dynamism is not often matched by parallel foresight and commitment towards protecting the safety and health of the production workers. Without governmental pressure and influence, the motivation for improvement in working conditions, in situations where trade unions are weak and jobs hard to find, is all too often lacking. Governments in turn, however,

*International Labour Organization.

require the stimulus of external guidance and it is here that the international labour standards adopted by the tripartite International Labour Organization play a pervasive role, all the more effective because the process of education and guidance is shared by workers and employers organizations as well as governments.

Nevertheless for a number of reasons industrial accidents and disabilities receive in some countries less attention that their effects deserve. Occupational medicine, for example, is often outside the mainstream of local medical research; factory safety is administered by, and accidents investigated by, the local Ministry of Labour (or equivalent) and not the Ministry of Health; the incidence of accident and disease is often under-stated or over-looked, as evidenced by the wide discrepancies that frequently exist between the number of accidents reported by employers and the number of claims for workmen's compensation. (This has been noted even in the United Kingdom with its long-established factory laws and inspectorate.)

Certain industrial diseases have long been recognized, and protective standards established, in industrialized countries — *byssinos* in the textile industry, *silicosis* in mining and quarrying, and *asbestosis* in processes using asbestos. But new dangers are continually revealing themselves, in particular substances used in industries processes which are carcinogenic in their effects. It was a piece of personal research by a doctor in the United Kingdom which revealed that one of the chemicals used in the production of PVC was a primary cause of bladder cancer among the workers at a particular factory. Although the dangers of asbestosis have been recognized for more than half a century, it is only in the past few years that it has been shown the dangers have been seriously under-estimated — in particular the dangers from blue asbestos (*crocidilite*). The increasing use of new forms of insecticide (and even fertilizers) in agriculture in developing countries extends the area of possible new dangers among workers completely unprepared for them — and in conditions where any ill-effects might not be quickly observed.

No such dramatic effects are possible in the industrial safety field as can result from, say, mass immunization. But there exist sound and tested international standards covering all aspects of the work environment which have been developed by the ILO — often in collaboration with other UN Agencies — in particular the WHO. There are few aspects of the safety, health, and welfare of workers which are not already covered by International Labour Conventions and Recommendations; model codes of practice including material for worker training in the identification of work risks. The poorer countries of the world are inevitably handicapped in the area of research and it is vital therefore that they have ready access to the experience of more industrially developed countries. The ILO's work in commissioning specialized research, and collating, analysing, and disseminating information largely fills this gap and provides a professional basis for improved safety standards and the reduction of avoidable accidents and disability. Direct assistance on the spot through the supply of high-quality experts has often had a marked effect on local standards and performance and is perhaps the quickest method of reducing industrial disablement and giving motivation and direction to local efforts, particularly when applied to specific occupations, such as coal mining, where the risks of accidents and disease are high.

In the final analysis, the real impact of international support of safety and health programmes at workplace must be measured against the extent to which national occupational safety and health programmes are developed and effectively implemented. What is needed is a greater political will in certain countries to adopt international standards; make the resources available for effective advisory and inspection services; use the technical and professional back-up now readily available from international sources, in particular the International Labour Organization in co-operation where appropriate with the World Health Organization; and ensure that tried and tested safeguards are adopted by employers and supported by worker representatives. On occasion much persuasion is needed to induce workers to adopt safety measures, sometimes because it reduces their earnings and sometimes merely because it is uncomfortable, e.g. the wearing of face masks in the tropics. The process of improving safety in workplace can never be regarded as static. It is a matter of constant vigilance to identify, evaluate, and control an ever more complex range of occupational hazards, some old and traditional, others new and often more deadly. Industrial injuries and industrial diseases can never be prevented altogether. But there is abundant evidence that much injury and disablement occurring at the workplace is both avoidable and preventable. The new industrial revolution has the resources to learn from the errors of the past, and to diminish the human misery and economic loss which is today reflected in the workplaces of the world and particularly in those countries where the processes of industrial production are a recent import.

4

The principal causes – chronic conditions

Leprosy

*Stanley Browne**

Introduction

While leprosy cannot vie in size with the major cripplers in the world today, its importance in terms of human tragedy and social consequences for both the victim and his family cannot be gainsaid or ignored. Today, on an island in the Far East, there are 2500 human beings suffering from paralysis and anaesthesia in all four limbs; 500 of them are also blind and thus are doubly cut off from the land of the living.

I am privileged to have worked in two situations as far as leprosy is concerned in the pattern of primary health care. In the former Belgian Congo, a church-based comprehensive primary health-care programme covering a large area was in existence before effective leprosy treatment became available. When the sulphones arrived, leprosy treatment and control were incorporated without difficulty into the health-care programme. The opposite situation obtained in Eastern Nigeria: a fully developed countrywide leprosy service was in process of being transformed into a polyvalent health service by in-service training and upgrading of leprosy workers so that they could thereafter function as public health auxiliaries working through all-purpose health centres.

Prevalence

There are probably 15 million people suffering from leprosy in the world today, of whom about a quarter have some degree of disability attributable to the disease. Since only about a fifth of those now suffering from active leprosy have access to treatment, a quarter of the remaining four-fifths, that is about three million people, can in the course of time be expected to develop disabilities because the progress of the untreated disease will result in preventable deformities that have not been prevented. And since leprosy is a chronic disease, symptomless in its onset and insidious in its progress, and characterized by deformities of hand, feet, and face, and slow impairment of vision leading to blindness — the personal tragedy and economic burden entailed by leprosy may not be sufficiently appreciated by governments. In addition, because the deformities of leprosy are hidden out of fear or shame, the true dimensions of the leprosy endemic in any country are frequently

*Secretary, International Leprosy Association.

unknown to the official statistical services or to the World Health Organization.

While very few countries are entirely free from leprosy, the main bulk of sufferers live within the medico-geographical tropics, with high prevalence rates generally in the sub-continent of India and South-East Asia generally and most of Africa (especially Nigeria and Zaire), and moderately high rates in Central and South America. India has the unenviable distinction of being the country with the highest estimated total of sufferers from leprosy — four million.

The precise reason or reasons for the disappearance of leprosy in North-Western Europe are quite unknown. It is likely that socio-economic factors (e.g. domestic overcrowding), played a significant part — as they certainly did in the more recent reduction in prevalence in Norway and Japan. Despite the arrival in the United Kingdom of 1140 people suffering from leprosy since 1951 (when the disease became notifiable), leprosy has not re-established itself in these islands: there have been no secondary cases. Leprosy has, however, persisted in the United States, particulariy in the southern States from California to Florida, and is still endemic in southern Europe — Portugal, Spain, Italy, Greece, Turkey, USSR, Malta, and Cyprus.

Since leprosy is a chronic disease and since the deformities that result from the untreated disease are progressive and cumulative, it is likely that in the absence of adequate control measures in most countries with a sizeable leprosy problem the actual numbers of those suffering from some degree of disability or handicap due to neglected leprosy will in the future increase in the world as a whole.

Main causes of disability in leprosy

The strictly medical causes are more easily assessable than the non-material, non-physical components. For ordinary purposes, leprosy can be considered as a slightly contagious disease caused by a specific micro-organism that produces disease only in human beings. The organism has a predilection for peripheral nerves and skin and the lining of the nose. By far the most important complications of leprosy are due to damage to the peripheral nerves, and hence to motor and sensory deficit in the tissues supplied, chiefly muscles and skin. Since these complications are in the main no essential part of the original infection, the results of nerve damage should in theory be amenable to preventive measures if the early infection is recognized and adequately treated before nerve damage has occurred. The claw hand, ulcerating fingers, ulcerating and drop feet, exposed sightless eyeballs — the conventional picture of leprosy — are no part of the original infection, but are the consequences of neglect.

The non-medical components of this horrifying result reside in the age-long prejudice of society, in many cultures and ethnic groups, to leprosy itself and to its victims. Ostracism, unimagined cruelties, and discrimination of all kinds — social, physical, religious, legal — are still practised against the unfortunate innocent victims of this slightly contagious infection. The social component of any public health programme is thus of equal importance to the strictly medical or therapeutic.

Programmes for prevention of deformity

The World Health Organization, through the Reports of its successive Expert Committees, makes recommendations to governments based on the latest information available. The governments adopt the Reports and recommendations at the General Assembly, and then decide for themselves how far they are able to implement them in their own countries. There is no co-ordinated international programme based on the WHO recommendations. In general, governments are left to themselves to formulate strategies for the control of leprosy itself and the rehabilitation of those suffering from the deformities associated with the disease. The unfortunate result is that leprosy is not generally accorded the degree of priority in official programmes that its importance and prevalence merit. This understandable reluctance on the part of governments is attributable to several mutually reinforcing factors, both medical and social: the inherent difficulties of detection and diagnosis; the long duration of treatment required and the slow clinical response to treatment; the considerable backlog of residual deformity; the chronicity of a disease that does not kill, and is never explosively epidemic; the strange aura of folklore and social attitudes surrounding leprosy, compounded of frank superstition and associated in many cultures with guilt and punishment.

A very practical aspect of policy in the prevention of disablement due to leprosy concerns the proportion of resources that might properly be devoted to the treatment of established deformity — and this in the light of the value of reconstructive surgery in reducing the stigma associated with specific deformities, in inducing more leprosy sufferers to present themselves for treatment at an earlier stage, and in showing that patients whose disablement is due to leprosy can be treated in units open to patients whose deformities are due to a variety of causes.

In general, and in most countries having a moderate backlog of patients with residual deformities due to neglected leprosy, it is reasonable to advise that not more than 10 per cent of the leprosy budget should be devoted to these recuperative activities. If this percentage is exceeded, it is more than likely that insufficient emphasis is placed upon the primary prevention of disablement by secondary prevention of the disease. Early diagnosis and adequate treatment is the best and finest method at present available for reducing the risk of nerve damage and for interrupting the cyle of transmission.

In many countries, and with many voluntary organizations, the photogenic appeal of 'before and after' pictures may be transiently convincing. A good anatomical result, however, for instance after tendon transplantation in the hand, takes no cognizance of the failure to restore lost sensation — impossible to show photographically.

Potential for prevention

The individual members of the International Leprosy Association, many of whom occupy positions of responsibility in their governments' health policy planning departments, are striving to ensure that the programmes for leprosy control are accorded adequate priority in programmes for the control of transmissible disease, in the realization that early diagnosis and adequate treatment of the active disease would eventually lead to a marked reduction in the incidence of deformity and disability.

Notwithstanding the availability of effective treatment for leprosy for over 30 years, the dimensions of the leprosy problem in the world show little sign of decreasing. In addition, there are two serious complicating factors that necessitate a radical rethinking of the strategy of leprosy control: the emergence of sulphone-resistance consequent on prolonged monotherapy necessary for a chronic mycobacterial disease; and the demonstration of 'persister' organisms, dormant and drug-sensitive, but unaffected by concentrations of drugs that should prevent bacterial multiplication. The result in either case is clinical relapse — either sulphone-resistant or drug-sensitive.

Apart from these measures, which constitute at present the most hopeful methods of control, research is proceeding apace in many co-operating laboratories throughout the world, under the stimulating aegis of the World Health Organization Programme for Research and Training in Tropical Diseases (especially its IMMLEP component), which is utilizing quantities of armadillo-derived leprosy organisms in the development of a specific, safe and effective vaccine. If and when this vaccine becomes available, it will be some time before it can be given to the millions in the Third World who would benefit from it. Meanwhile large-scale chemoprophylaxis — while reportedly protective in about three-quarters of contacts — is open to criticism on grounds mainly of cost and difficulties of supervision of regular administration of a toxic drug over long periods to people exposed to, but not actually suffering from, the disease itself. Patient compliance is low; contact compliance to a prolonged course of prophylactic drug would probably be lower.

Vaccine

In the search for a protective vaccine, much work has been done in many directions. The real difficulty lies in the fact that resistance to leprosy does not depend on the elaboration of specific humoral antibodies, but on the specific sensitization of lymphocytes to recognize and to lyse invading *M. leprae* — i.e. it is a cell-mediated immunity. In point of fact, in the form of leprosy occurring in subjects who can mount no cell-mediated immunity at all, there is a plethora of several kinds of antibody in the serum, but instead of limiting the spread and multiplication of the invading organisms, they may actually damage target organs such as the peripheral nerves and the uveal tract of the eye. Thus, many kinds of antibody (gamma-globulins, immune complexes, and complement), are produced in quantity in response to the presence of multiplying organisms, but they serve no useful purpose. The search for something that will specifically sensitize lymphocytes 'to do what they can't do, or won't do' encounters unexpected difficulties and setbacks, and is likely to be protracted.

Recent work has concentrated on identifying potentially useful moities of *M. leprae* as antigens, and on potentiating BCG with extracts of *M. leprae*, *M. tuberculosis* (avian), *M. vaccae*, etc.

Well-documented field trials of BCG have yielded discordant results. The differing degrees of leprogenicity and the influence of anonymous mycobacteria are complicating factors.

If and when a possible vaccine becomes available for large-scale trial, the logistic and ethical problems will loom large. Comparable populations should in theory continue to be exposed to leprosy; one-half should receive the possibly protective vaccine — the other half, not. Whether it is ethically justifiable to withold treatment from index cases suffering from leprosy in both groups, so that exposure to risk may be comparable, is a nice point.

Another point is the possible damage to peripheral nerves if (as the result of the vaccination) cell-mediated immunity is enhanced in individuals with early mycobacterial invasion of nerves.

Several years will necessarily elapse before it can be shown that a vaccine affords real protection against leprosy. And then there will come the problems of administration — single or multiple doses, optimum age, contra-indications.

In the absence of primary prevention by a specific vaccine, or chemotherapy, or enhancement of resistance by immunotherapy, the control of leprosy at present consists of secondary prevention by reducing the source of living leprosy organisms in the vicinity of susceptible contacts.

Programmes of rehabilitation of those suffering from the deformities associated with leprosy

If governments generally accord a low priority to the control of leprosy within the context of the control of transmissible disease, they accord a still lower priority to the recuperation of people suffering from the deformities of hands, feet, and face commonly attributed to leprosy, leaving this task largely to voluntary agencies. Voluntary agencies (mostly Christian missions) were the pioneers in caring for and treating sufferers from leprosy. Now united under the aegis of ILEP (the International Federation of Antileprosy Associations), they are responsible for the treatment of about one-third of all leprosy sufferers in the world who are actually getting treatment for their disease.

ILEP members raise annually about 32 million dollars for leprosy programmes in the Third World. Fund-raisers are being gradually weaned by medical arguments away from photogenic appeals that rely on horrific pictures of deformity and the results of reconstructive surgery, and towards a greater emphasis on the need to prevent preventable deformity by good treatment of patients diagnosed at an early stage of the disease.

The cost and difficulties of reconstructive and plastic surgery, physiotherapy, vocational training, job placement, sheltered workshops, and home industries, have to be seen in the light of the social stigma of deformity due to leprosy and the reluctance of many communities to accept patients who have had leprosy and whose deformities have been surgically corrected. Added to this is the impossibility of operatively restoring lost sensation. In a rural community, an agricultural worker may be accepted, but a town-dwelling artisan meets with hostility and rejection.

Some few centres have adopted an integrated approach, where leprosy sufferers share with those whose disability arises from other causes, in facilities for reconstructive surgery or job training or sheltered workshops, but such centres are few and far between. Similarly, leprosy sufferers usually have to take a back seat when it is a question of the provision of prostheses and appliances.

Programmes for prevention of deformity

It is unfortunately still true that the political will is lacking in many such countries, and that agreement with forward-looking policies expressed at Geneva does not necessarily imply advocacy of unpopular policies back home. In addition, recruitment of doctors and other staff into a leprosy service is hampered by such factors as the supposed segregation of leprosy workers; the linguistic and cultural associations of the disease; and the diminished opportunities for private practice in leprosy. However, many national doctors are now attracted by the research openings in leprosy, especially in the fields of immunology and microbiology.

There are certain positive advantages presented by leprosy as a transmissible disease when its control is contemplated. These are: the absence of an animal reservoir; the relative ease of identification of the disseminator of viable leprosy organisms; the importance of person-to-person contact, without the interposition of an insect vector; and the extraordinary susceptibility of *M. leprae* to minute concentrations of dapsone.

The supreme needs for the control of leprosy in the world today are:

1. Education in its broadest sense of medical personnel (including students, doctors, dermatologists, community health physicians), of politicians and opinion-formers (in the mass media, in schools, etc.).

2. Definite training, appropriate and practical, of all medical and paramedical staff likely to come into contact with leprosy, so that the whole body of erroneous assumptions that leprosy is different may be destroyed.

3. Research specifically directed towards (i) the culture of *M. leprae* in artificial media; (ii) the production of an effective, specific, and safe vaccine that can be produced cheaply and given by auxiliary workers; and (iii) the discovery of new drugs that are cheap, safe, and specific.

When leprosy is controlled, disablement due to neglected leprosy will be a thing of the past.

Suggestions

In view of these diverse social and medical problems surrounding the complex issue of the prevention of deformity due to leprosy in the world, there is no easy or simple solution, but I would be so bold as to make the following suggestions:

1. To ensure, by advocacy and financial support, that leprosy forms as essential part of the World Health Organization programme for the implementation of the Alma Ata Declaration. 'Health for all by the year 2000' will prove to be an empty and meaningless catchphrase unless the back-up in training, supervision, and medical supplies necessary to the programme is forthcoming.

2. To encourage and support, in countries facing a forgotten or neglected leprosy problem, pilot schemes in which leprosy is integrated into programmes for the control of locally important endemic diseases (like tuberculosis, malaria, schistosomiasis, onchocerciasis) and family planning.

3. To support voluntary agencies engaged in co-operative efforts with governments in programmes for leprosy control that include health education, integrated treatment schemes, research into drug regimens,

and training programmes especially directed towards the control of leprosy in such countries.

4. To initiate rehabilitation services in which those handicapped by deformities due to leprosy receive expert assistance (physiotherapy, reconstructive surgery, vocational training, etc.) together with those whose deformities are due to congenital abnormalities or infections (e.g. tuberculosis, poliomyelitis) or trauma (accidents — traffic, domestic, factory, or mine). The evidential value of reconstructive surgery has an incidental advertising impact on leprosy control programmes.

5. To encourage the making of simple prostheses, using local materials and available skills in producing acceptable, repairable, cheap, and non-stigmatizing appliances, footwear, etc., for patients suffering from deformities, whatever the cause.

6. To stimulate the efforts of the Medical Research Council, Leprosy Section, in its co-operative efforts to elaborate a specific, safe, and effective vaccine against leprosy and to support the activities of the IMMLEP and THELEP projects of the WHO Programme for Research and Training in Tropical Diseases (Leprosy) to this end.

If these suggestions can be adopted and implemented, not only will the stigma still surrounding leprosy be undermined, but the future incidence of deformity due to leprosy will be reduced to manageable dimensions.

Tuberculosis: a major clinical, therapeutic, and rehabilitative problem in developing countries

Balu Sankaran* and A. Piot†

The problem

Tuberculosis, which is a rather forgotten entity in almost all countries of the developed world, is still a major public health problem in most developing countries. In terms of 'human suffering' the sum total of the individual suffering caused by the disease and the related social costs is appalling.

In epidemiological terms *the incidence of new cases* found in a specific year would be a good index. This would indicate the extent of susceptibility of the population groups and the trend of the problem. This index unfortunately is of value only in countries where the coverage of the health system is high, the quality of the diagnostic methods is good, and the notification mechanisms are reliable. In practically all developing countries this is not so.

*Director, Division of Diagnostic, Therapeutic and Rehabilitative Technology, WHO Headquarters, Geneva.
†Chief, Tuberculosis and Respiratory Infections Unit, WHO Headquarters, Geneva.

The *prevalence of different categories of cases of pulmonary* tuberculosis has to be determined in comprehensive surveys. Such surveys are, however, difficult to carry out and quite costly. Technically, it is not a good epidemiological indicator to measure the trend of the problem, although it gives useful information for programme planning and evaluation.

The annual *risk of infection* is the best single indicator that gives an estimate of the proportion of population which will be primarily infected or reinfected with tubercle bacilli in the course of one year. It is derived from the results of tuberculin testing in a representative sample of unvaccinated children. However, its determination in developing countries is greatly hampered by the widespread immunization with BCG among young children and the high prevalence of non-specific tuberculin sensitivity. In spite of these drawbacks there are certain conclusions that are warranted: (i) In developing countries the prevalence is definitively much higher than in developed countries; (ii) there is no perceptible decline in the incidence of disease in many developing countries, a decline that had set in a long time ago in all developed countries; (iii) it is mainly a disease of the young adults in the developing countries affecting the wage earner in many instances, but it is also present in the elderly; (iv) in most developing countries the risk of infection is of the order of 2–5 per cent, about 20–50 times higher than in technically advanced countries. The gap has been constantly increasing, because in developed countries the risk of infection decreases by 10–12 per cent a year, whereas in many developing countries the risk has remained unchanged for many years or has only been declining very slowly; (v) there is a relative constant ratio (lower for developed and higher for developing countries) between the annual risk of infection and the incidence of smear-positive pulmonary tuberculosis, namely, 1 per cent of annual risk of infection corresponds to about 50 new cases of smear-positive pulmonary tuberculosis a year per 100 000 population. Hence by a conservative estimate there should be four to five million new cases of smear-positive pulmonary tuberculosis each year in the world, or about ten million if all forms of pulmonary and extra-pulmonary tuberculosis are taken into account.

Control methods

BCG VACCINATION

Although several controlled trials produced contradictory results on the effectiveness of BCG vaccination, no negative results have ever been reported from studies conducted among infants and young children. Present WHO policy recommends BCG vaccination of the newborn or at the earliest possible time in developing countries to prevent childhood tuberculosis, especially the serious miliary form and meningitis.

The effectiveness is the product of the actual efficacy — 80 per cent in the best conditions — by the vaccination coverage. From an operational point of view BCG vaccination is a component of the Expanded Programme on Immunization (EPI). Freeze-dried vaccines are preferred because of their stability at room temperature even in tropical climates. The intradermal method is recommended because it is the only one which ensures a uniform dose.

A trial carried out in South India by a very careful assessment showed a complete absence of a protective effect during a follow-up of seven and a half years against bacteriologically proved pulmonary tuberculosis in adolescents and older age groups (young children, although included in the trial, were not observed). There is no suggestion that the trial methodology, nutrition, or quality of the vaccine used had been responsible for the negative result. Instead, other factors have been considered as possible causes that reduced the efficacy of BCG. Among them the high prevalence of environmental mycobacteria and of leprosy in the trial area; the high proportion of disease caused by the South Indian variant of the tubercle bacilli, and the unusual epidemiological pattern of tuberculosis in the trial — the main characteristic of which was the long delay between the moment of the infection and the start of the disease. Also the tuberculin sensitivity induced by BCG vaccination in the trial population appeared to have waned considerably within a few years. Further research is needed to identify with certainty the factors that modify the effect of BCG vaccination especially in tropical areas.

PREVENTIVE TREATMENT (CHEMOPROPHYLAXIS)

Mass preventive treatment is not feasible and hence is not recommended in a community health programme. In addition to the operational difficulties and economic impossibility, there are also some toxic effects induced by isoniazid, especially hepatotoxicity, which do not favour its widespread use in the general population as a chemoprophylactic drug. Its only place is probably in some high-risk groups, such as children, contacts of newly diagnosed infectious cases of tuberculosis.

CASE-FINDING AND TREATMENT

The most practical way to produce a lasting impact on the tuberculosis programme is through case-finding and treatment, namely to detect the sources of infection as early as possible and render them non-infectious by chemotherapy. It results in an immediate reduction in the prevalence of smear-positive cases of pulmonary tuberculosis and a self-perpetuating decline of the problem.

Mass miniature radiography has proved inadequate screening procedure for tuberculosis. It is very expensive, and contributes only a small proportion of the total number of cases reported by the health services. Other disadvantages for developing countries are that it requires the services of highly qualified technicians and medical staff and there are frequent mechanical breakdowns of the equipment because of poor maintenance when working in rural and inaccessible areas.

Bacteriology should play the main role in detecting new cases of tuberculosis. Direct smear examination of sputum of patients with respiratory symptoms is the basic technique to be applied through the extension of laboratory services to the level of primary health care. Examination of direct smears is relatively simple, inexpensive, and detects those cases of pulmonary tuberculosis which are the most infectious.

Adequate chemotherapy should be given free of charge to every patient with infectious pulmonary tuberculosis. Hospitalization is not required in the vast majority of cases. Regimens of one year's duration are still widely used in developing countries. The two most common regimens are: (i) isoniazid plus thiacetazone in daily self-administration, often with a supplement of streptomycin during the first month of treatment; (ii) isoniazid plus streptomycin, twice a week, fully supervised by the health service.

The success rate of chemotherapy under programme conditions is variable, but not higher than 60–65 per cent in the best circumstances. In many instances the success is much lower, with defaulter rates of up to 50 per cent. Default and irregularity in the self-administration of drugs is outstandingly the most important reason for failure of treatment. Supervision of drugs intake has been strongly recommended but there have been many difficulties to organize it efficiently. Other problems are the non-availability of drugs at the primary health care level and the frequent movement of patients during their treatment from one area to another.

There are now a number of short-course regimens of six to nine months' duration that are highly effective, of low toxicity and well tolerated. These regimens are based on an initial phase of isoniazid, rifampicin, and pyrazinamide supplemented by a fourth drug (streptomycin or ethambutol), with several alternatives in the continuation phase. Some developing countries are already using short-course regimens in their national tuberculosis programme. However, the cost is still an obstacle to make widely available short-course regimens for all developing countries. The great advantage of these regimens is that a high proportion of patients are cured even within the first three months of treatment, so that they offer an important degree of protection against premature default from treatment.

When the disease has reached an advanced stage both in the pulmonary and extrapulmonary lesions, in some cases surgical intervention may be indicated. Surgery for tuberculosis has still a place in developing countries. Tuberculous paraplegia requires transthoracic decompression and fusion of the involved vertebral bodies and fusion of the cervical spine in unstable pathological dislocations. Similarly abdominal and kidney lesions, including tuberculous pachymeningitis and tuberculomas of the brain and spinal cord are still lesions that require surgical treatment.

National tuberculosis programme

All the activities to control tuberculosis should be co-ordinated within the framework of a national tuberculosis programme. The programme in all countries needs strong technical support and supervision initiated and promoted by the Ministry of Health. At intermediate level, special managerial teams may be desirable, but they should form part of the general health-care system. All the control activities should be integrated into the primary health-care level, as a permanent service, organized on a country-wide basis.

Training of personnel is vital. Case-finding and treatment will be effective only if carried out by suitably trained staff at the primary health-care level and technicians at the laboratory network. There should be a systematic supervision of the personnel, as a continuation of the training. The regular supply of drugs and materials for the microscopic examinations are essential for the maintenance of the programme.

Monitoring and evaluation of the programme should be based on a simple but meaningful recording and reporting system. The programme should be implemented with community participation. Its involvement requires education activities to make the population aware of the symptoms of tuberculosis and motivate the individuals with symptoms to seek health care earlier and collaborate with treatment.

A disease that can be and should be controlled within the span of a few decades is worthwhile making a commitment to, and a sustained effort for eradication is a must to avoid much human suffering: ultimately disability from tuberculosis can and should be prevented.

Stroke and hypertension

Henry H. Betts*

The United States has developed a high degree of awareness concerning the national health problem of stroke and hypertension. Stroke is the third cause of death in the United States. At least 200 000 Americans die annually from stroke. It is believed that one American in ten will die of stroke.

*Magnuson Professor and Chairman, Department of Rehabilitation Medicine, Northwestern University, and Executive Vice-President and Medical Director, Rehabilitation Institute of Chicago, Illinois, USA.

Another important factor is that 300 000 Americans *survive* stroke each year. Of these, 70 per cent are disabled in a way that interferes with employment, mobility, or social life (Framingham study). According to the National Institutes of Neurological and Communicative Disorders in Stroke, there are 2.5 million stroke survivors presently in the United States — 30 per cent have returned to work or normal lifestyles; 55 per cent are partially disabled and can more or less care for themselves; and 15 per cent must stay permanently in a nursing home.

Cerebral thrombosis is the cause for 63 per cent of strokes (Framingham study). One half of those patients who live are alive five years later. Of these patients, 21 per cent are unable to care for themselves. One in five is alive and functioning on his own after eight years.

Cerebral embolism causes 15 per cent of strokes (80 per cent of cerebral emboli are associated with heart disease). Cerebral embolism kills 20–40 per cent of its victims within three months.

Subarachnoid haemorrhage kills one-half of the patients within a few days (in one study, the death rate was 80 per cent). The chance of dying of a second haemorrhage within six months is 20 per cent. Intracerebral haemorrhage mortality is 90 per cent. Those who die of intracerebral haemorrhage survive more than one week after onset (87 per cent Evanston Hospital, Illinois). Patients who die of subarachnoid haemorrhage succumb in seven days or less (90 per cent).

Awareness and continuing education to the many risk factors can aid in the prevention of stroke and disability. Risk rises sharply in later years.

There is no proof that family history is a significant factor; however, a few studies have found that identical twins are more likely to suffer a stroke than fraternal twins.

In European patients, prior heart disease may be an even more important risk factor than hypertension (Dr Gershon Librach, Tel Aviv, American Joint Distribution Comm., Malben Services). They show the incidence of stroke in patients with heart disease is nearly six times that of patients without known heart ailments. The incidence of stroke was highest in those with previous infarction, heart failure, atrial fibrillation, arrhythmia, and valve defects. Those with rheumatic heart disease and artificial heart valves risk a new source of embolic strokes.

Patients fed a diet high on unsaturated fats which tends to avoid high lipid levels were 41 per cent less likely to have a stroke than those on normal diets.

Diabetics have a greater risk of suffering a stroke due to the increased tendency to develop atherosclerosis, hypertension, and high levels of blood lipids than the general public.

It is believed that smoking may play an important role in the incidence of stroke. There are three times as many strokes among smokers. Smoking increases the incidence of ischaemic heart disease, thus, a known risk factor (Framingham study).

Some people feel that all patients with intracerebral haemorrhage may have hypertension whether it had been recognized or not. *Hypertension is the major risk factor for stroke*. It is 17 times more potent than any other risk factor (Framingham study). Hypertension is one of the three major factors that accelerates atherosclerosis causing one and a quarter million heart attacks and 650 000 heart attack deaths per year. Strokes of all types are 2–4 times as common in patients with pathological high blood pressure. A history of hypertension is found in 60–90 per cent of intracerebral or subarachnoid haemorrhage patients. At least 40 per cent of patients with untreated hypertension die of cerebral vascular accidents. Of patients with mild hypertension, 15–25 per cent will have non-fatal strokes (half are caused by intracranial haemorrhage). Figures show that 70 per cent of haemorrhagic stroke patients were found to be hypertensive (Harvard Stroke Registry); 59 per cent of non-haemorrhagic stroke patients were hypertensive (Harvard Stroke Registry). Statistics indicate that 40 per cent of patients with haemorrhagic stroke and 52 per cent of others have a past history of hypertension (Harvard Stroke Registry).

It is estimated that if all 23 million adults who have high blood pressure in the United States were screened and treated, the overall death rate from strokes (and heart attacks) would decline by 20 per cent.

Hypertension is the factor in 75 per cent of all first strokes. Of 400 000 people who have a first stroke each year, 54 per cent have definite underlying hypertension; 21 per cent have borderline hypertension, and 25 per cent do not have hypertension. Hypertensive men have ten times the risk of stroke as men with normal levels.

The economic cost of stroke in the United States is $4 605 000 000.

Hypertension

Hypertension is involved with 75 per cent of first strokes. Hypertension was the factor in 160 000 of the 214 650 stroke deaths in 1973. Thirty-two million Americans (one in six) have definite hypertension. Twenty-five million Americans have 'borderline' hypertension, a major contributor to 500 000 cases of stroke a year. People with hypertension are seven times more susceptible to stroke (three times more susceptible to coronary heart disease).

Eighty-six per cent of first stroke patients in the 45 to 74 age group have hypertension. Prevalence increases with age — 50 per cent greater among blacks than whites; higher for men than for women. According to the Framingham study, in both men and women, blood pressure proved to be the most reliable factor for predicting subsequent stroke. Hypertension was associated with a 5–30-fold increase in the risk of stroke depending on the subject's age or sex (hypertension equaled 160/95 or above). The risk of stroke was directly related to both systolic and diastolic pressures (this was confirmed by the Chicago Stroke Study).

In 1952, less than 50 per cent of Americans with high blood pressure knew that they had it. Less than one in eight were under treatment. All the evidence is resounding in favour of the strong relationship between hypertension and stroke.

There are few diseases that are simpler to diagnose than hypertension and the treatment of it is now possible by means of drugs. The issue becomes one of educating the population to have their blood pressure taken, offering them medication, and making it economically feasible for the patient to afford the pills.

The US approach to hypertension

The National Heart, Lung and Blood Pressure Institute is one of eleven Institutes, Bureaux, or Divisions in the National Institutes of Health. Its primary mission is to support biomedical research. In 1972, it established the National Heart, Lung and Blood Act indicating the acquisition of new knowledge and basic research and the responsibilities of the Institute — that it must institute clinical trials, validate activities, do technological assessments, train, and provide education.

In relation to hypertension, in the year 1980, $78 million was spent — about 17 per cent of the Institute's budget. The primary goal is prevention.

The most visible activity and dissemination by the Institute in the area of hypertension is through the National High Blood Pressure Education Programme. It was begun in 1972 with emphasis on awareness. It represents a national programme with all the federal agencies, volunteer agencies, private, and public sectors participating with a lead responsibility for the programme in the federal sector given to the National Heart, Lung and Blood Institute.

The programme was very simplistic. 'It was an education programme aimed at the professional and the general public to emphasize the high prevalence of hypertension; that it could have no symptoms, that it was easy to detect, that therapy was available, and that adherence to therapy was essential if one was going to have control and prevent the sequelae of the disease — namely, stroke, heart failure, and renal failure.'

Since the aim of the National High Blood Pressure Education Programme was to increase the number of people who have annual blood pressure checks, it was felt that the only way to ultimately ensure control was to have the programmes in place at the community level. To have a programme on the national level without its roots in the community would be to have a programme that was ineffective.

In professional education, the goals were to increase professional awareness of high blood pressure as a problem, to achieve wide adoption of standards for detection and therapy, to increase the involvement of the physician in patient education, to try to make these activities more effective, and to try to increase the team approach to high blood pressure control. Physicians were not attending to the problem of hypertension sufficiently.

The National High Blood Pressure Education Programme is a programme in which both the Federal and Private Sector have participated. The federal activities have been co-ordinated through an interagency technical committee. The private activities have been co-ordinated through the National High Blood Pressure Co-ordinating Committee.

The National High Blood Pressure Education Programme involves at least 15 federal agencies, at least 150 major national organizations with over 2000 organized communities participating. The programme began with the co-ordinating office in the office of the Director of the Heart, Lung and Blood Institute with a very large Education Information Advisory Committee that was chartered at the level of the Secretary of Health, Education, and Welfare. Four task forces were set up.

In July 1975, the Kidney Foundation, The Heart Association, The Citizens for the Treatment of High Blood Pressure, Inc., and The National High Blood Pressure Co-ordinating Committee were created to co-ordinate this national activity and to exchange information. Member organizations meet three or four times a year taking on the responsibility for sponsoring activities collaboratively or jointly, exchanging views, analysing problems, and developing plans. It was assumed from the beginning that the Coordinating Committee would provide a forum so that each of the organizations could present its priorities for all of the membership to look at and then identify any gaps or overlaps. The Coordinating Committee has had several functions: exchange of views, facilitiation of collaboration, and 'consensus building'.

Results

Stroke rates have been declining at the rate of about 1.5 per cent per year for the period of the sixties.

When the hypertension programme began, we saw a sharp break in the curve. In 1972, the rate began to decline at 5 per cent a year, and this has persisted through 1978.

Since the creation of the National High Blood Pressure Education Programme in 1972, there has been a 38 per cent decline in death due to stroke in the United States. (There has been a decline in deaths due to disease of the heart of 28 per cent). The total annual cost of the High Blood Pressure Control Programme is $90 per person and the cost of a stroke in the United States is estimated at $15 000.

Treatment of mild or borderline high blood pressure can reduce premature deaths by 20 per cent. A 17 per cent decline in mortality can occur if hypertensives are given care. When applied to mild hypertension, this can cause a mortality decline of about 20 per cent. Black citizens can benefit as much as whites. There is a 22.4 per cent decline in blacks with a 10 per cent decline of mortality in whites. Substantial benefits resulted from systematic treatment for those with diastolic pressures as low as 90.

In the Australian Therapeutic Trials (ATT), 3232 subjects were studied who had diastolic pressures between 95 and 109. There was a significant reduction in both non-fatal complication and mortality when treated and untreated patients were compared applying to both men and women both above and below 50 years of age.

A study from Gothenberg based upon patients with moderate hypertension diagnosed on community screening demonstrated a substantial reduction in the incidence of myocardial infarction when treatment was begun.

In the future, increased recognition and in-depth studies of hypertension and stroke worldwide are necessary. No countries, especially those with struggling economic and social problems, can afford the gigantic human strife and the cost of large populations of stroke patients. Chronic illness and disability are tolls humans should no longer suffer and have to support unnecessarily. Prevention throughout the world is the only solution to lessen this burden for all of us; and many lessons can be learned from the American experience.

Coronary heart disease

WHO Expert Committee

Despite medical advances, coronary heart disease (CHD) remains the leading cause of death in developed countries. Rates differ widely. In some countries (e.g. Belgium, Japan, United States) they are falling. Elsewhere (e.g. in several European countries and also in some developing countries) rates are rising rapidly. These differences and changes reflect *inter alia* differences in life-style and thus indicate the potential for prevention.

These concerns led the World Health Organization to assemble a group of leading cardiologists and epidemiologists from 12 countries. They met in Geneva from 30 November to 8 December 1981, and their report and recommendations on the prevention of CHD have now been published.* Whilst recognizing some important areas of uncertainty, the

*Prevention of Coronary Heart Disease. A report of a WHO Expert Committee. *WHO Technical Report Series* No. 678, Geneva (1982).

Committee was able to identify several preventive measures where the balance of evidence indicates a sufficient assurance of safety and a sufficient probability of benefit to warrant action now. Evidence of similar strength has in the past led to other major decisions on public health policy such as those on air pollution control, sanitary improvements, and the formulation of requirements for a healthy diet. Public health policy must be based on best-informed judgements.

Call for a population approach

The report emphasizes the necessity for changes in whole populations. This is a mass disease and therefore a *population approach* to prevention is essential. In high-incidence countries the levels of major risk factors are too high in most people; and most cases of CHD occur among the many people with 'average' risk, not among the smaller number with exceptional risk characteristics (such as very high blood cholesterol or blood pressure, or diabetes). What is required is a shift towards normality in the whole distribution and averages of risk factors. Even relatively small changes in whole populations could bring large benefits.

The relationships between habitual diet, blood cholesterol levels, and CHD are judged to be causal. High-incidence populations are advised, through progressive changes in eating patterns, to reduce their average levels of blood cholesterol in the direction of those found in countries where CHD is uncommon. A lowered intake of saturated* fats is particularly stressed, with a greater proportion of energy derived from vegetable sources.

Recent findings suggest that polyunsaturated fats may reduce the risk of thrombosis; nevertheless, present evidence does not justify general advice to *increase* the intake of polyunsaturated fats (although mono- and polyunsaturated fats may be used to offset some of the reduction in saturates).

Salt and smoking

A progressive reduction in average salt intake might produce some

---

Saturated fat: a fat so constituted chemically that it is not capable of absorbing any more hydrogen. These are usually the solid fats of animal origin such as the fats in milk, butter, meat, etc. A diet high in saturated fact content tends to increase the amount of cholesterol in the blood. Sometimes these fats are restricted in the diet in an effort to lessen the hazard of fatty deposits in the blood vessels. *Unsaturated fat*: a fat whose molecules have one or more double bonds, so that it is capable of absorbing more hydrogen. Monounsaturated fats, such as olive oil, have only one double bond (the rest are single) and seem to have little effect on blood cholesterol. Polyunsaturated fats, such as corn oil and safflower oil, have two or more double bonds per molecule and tend to lower blood cholesterol.

lowering of average blood pressure, and thus bring about a safe, cheap, and substantial reduction in CHD and in the need for antihypertensive medication. Blood pressure reduction may also be assisted by control of obesity and excess alcohol intake in the population.

Smoking contributes importantly to CHD, and non-smoking must come to be regarded as normal behaviour. As far as CHD is concerned, present evidence does not support promotion of the so-called 'safer cigarette'.

Population obesity relates to an unduly low average energy expenditure. Regular exercise may help to reduce obesity and other important risk factors.

The WHO report stresses the need for national strategies to achieve these objectives, starting with young people. All sections of the community need to be involved — medical services, government, schools, community organizations, agriculture, the food industry, and the mass media. Every country should have a plan of action.

Information systems are needed: each country should know what changes are occurring in CHD mortality and incidence, national diet, and population levels of risk factors.

More knowledge is also needed on how to improve public awareness and understanding and on the influences and constraints governing relevant changes in behaviour.

Finally, the report stresses that in developing countries CHD threatens to extend as socioeconomic development progresses. Policies on nutrition, control of smoking, and avoidance of sedentary living and obesity are essential to prevent the development of the patterns of risk factors that are familiar in developed countries and which would lead to mass CHD.

The age factor

The WHO Regional Office for Europe

Current demographic projections indicate that by the year 2000 approximately 20 per cent of the European population will be 65 years of age and over. The elderly are one of the fastest growing subgroups in the population, particularly the very old, i.e. those aged 80 years and over. The degree of disability increases with advancing age. Women, in general, report a higher level of unmet social needs than men.

Providing services for the disabled elderly requires the development of an adequate information base. Until now, information on the incidence and prevalence of disability was only partly available from epidemiological studies.

The prevalence of disability among the elderly

It is difficult to assess the prevalence of disability in elderly groups, because the physiology of aging is only partly understood, and it is difficult to distinguish it from pathological conditions. Elderly individuals commonly have multiple disabilities, which interact and influence their performance in a qualitatively different way than could be expected from individual disabilities themselves. Clarification of the concepts of illness, impairment, disability and handicap has been published only recently on an experimental basis[1] and is not yet in everyday use.

The term 'impairment' refers to dysfunction of bodily parts or organs. Disability is a characteristic of an individual, describing aberrations in normal performance, whether physical, emotional, mental, or social. Handicap is the social consequence of disability, i.e. the disadvantage compared with other individuals.

Epidemiological studies on the prevalence of disability in the elderly have not used standard terminology and suffer from all the 'teething troubles' mentioned above. However, the prevalence of disability and the need for social and medical support for the elderly have been investigated in many countries. Severe disabilities due to disease do not differ considerably in number from figures estimated in middle age.

In a study in Gothenburg on 70-year-olds, only 3 per cent were found to have disabilities requiring care in an institution, such as a hospital, nursing home or home for the elderly.[2] It should be emphasized that about one-half of the disabled had been cared for in institutions for decades while the other half did not acquire their disease and/or handicap until late in life. At the age of 75 years, an additional 2 per cent of the population in Gothenburg needed institutional care.

The indications for institutional care obviously vary from country to country, and many disabled persons are, in fact, being cared for at home. Thus, in the Gothenburg study, 2 per cent of the 70-year-olds and 4 per cent of the 75-year-olds living at home were found to be severely disabled. In certain areas of Scotland, it was found[3] that for every severely disabled old person in hospital, there were two with a similar degree of disability at home; for every two incontinent persons in hospital, there were three at home; and for every three with severe mental abnormality in hospital, there were four at home. The results indicate that substantially more disabled old people are cared for at home than in institutions in the United Kingdom. These prevalence figures might well have been lower had more effort been applied to prevention earlier in life.

As to the physical ability of old people living at home, the following observations have been made. In the Gothenburg study, 18 per cent of the 70-year-olds declared that they needed some sort of assistance at home, e.g. with cleaning of windows and floors, laundry and other heavier activities. At the age of 75 years, the corresponding figure was 31 per cent. In Scotland, the problems of disability in the elderly and social needs were investigated in 1975,[4] when it was found that 58 per cent of 762 interviewed people aged 65 years and over needed help or support from a social worker and 11 per cent were in need of domestic help.

In a Finnish study, Kalimo *et al.*[5] found 50 per cent of those aged 65–75 years and 30 per cent of those 75 years and over to be totally without mobility impairments and, in addition, about 25 per cent in each age group to be able to move around outdoors without assistance by another person despite modest impairment.

To sum up, available data indicate that severe disability occurs in 3 per cent of 70-year-olds and in 5 per cent of 75-year-olds to a degree that makes institutional care necessary. Milder forms of disability were found in another 2 per cent of the 70-year-olds and in 4 per cent of the 75-year-olds, e.g. loss of the ability to perform hygienic functions. The frequency of these disabilities in higher age groups remains to be investigated. The elderly need help to perform heavier housekeeping tasks to a greater extent than do younger individuals, but knowledge of the needs of those who are older than 75 years is not well established.

TRAFFIC ACCIDENTS

In many countries, traffic accidents are common among the elderly. In the Scandinavian countries the mortlity rate from motor vehicle accidents increases with age from 40 to 80 years of age. In Sweden, the number of elderly car drivers and passengers injured or killed in traffic accidents increased by more than 50 per cent for drivers and by 38 per cent for passengers from 1970 to 1979. However, most traffic accidents happen when the old person is walking or riding a bicycle. In the Gothenburg study, the incidence of fatal accidents was found to be about four times higher in an elderly group (65+) than in a group aged 5–14 years. Similarly, the proportion of severe injuries in connection with walking was 1.5:1, but mild injuries occurred with the same frequency. Injuries to the elderly require longer rehabilitation, and the elderly will often die from injuries that younger people will survive. Factors that influence the performance of the elderly in traffic include changes in the function of sensory organs and loss of agility.

In the Gothenburg study, it was shown that only a limited number of persons among the 70-year-olds and, in another study, no women and only 20 per cent of healthy men aged 80 years[6] were able to walk at a speed of 1.4 metres per second, which is necessary to cross the street at intersections regulated by traffic lights in the city of Gothenburg.

PREVALENCE OF SOMATIC DISEASES

Recent epidemiological studies have improved the possibility of differentiating between the manifestations of aging and symptoms of disease at more advanced ages. The longitudinal population study in Gothenburg also demonstrated that both 'underdiagnosis' and 'overdiagnosis' are common in the elderly. The elderly themselves are often convinced that certain somatic and mental manifestations are due to aging and, therefore, do not mention these to the doctor. Typical examples are problems with urinary incontinence and mental depression. Another reason for underdiagnosis is that the symptomatology is often more vague or different in the elderly compared to other individuals. The Gothenburg study showed, however, that overdiagnosis is even more common, meaning that the manifestations of aging are interpreted by physicians as symptoms of disease. The supposed prevalence of hypertension and of diabetes in the elderly illustrates this problem. These are areas where adequate diagnostic criteria are lacking at present.

Studies have shown, for example, that shortness of breath, slightly higher blood pressure and higher heart volume may be normal at the ages of 70 and 75 years. Yet this is commonly taken by doctors as evidence of cardiac failure and treated as a disease. In the Gothenburg study, 13 per cent of 70-year-old men and 20 per cent of 70-year-old women felt subjective shortness of breath (dyspnoea) that was not statistically correlated to any definable disease. As far as can be judged from the longitudinal follow-up to the study, fewer than 50 per cent of those treated with digitalis drugs have shown any evidence of cardiac disease that could be expected to react favourably to such treatment. Dyspnoea is obviously common in the elderly but need not be caused by disease. In an epidemiological study in Finland, Sourander *et al.*[7] found that 34 per cent of the population over 65 years of age complained of shortness of breath compared with 46 per cent in Gothenburg. It is thus extremely difficult to present any reliable figures on the prevalence of definable heart disease at higher ages, as it is obvious that symptoms considered to be characteristic for cardiac diseases are also common as manifestations of physiological aging.

It is difficult to assess the prevalence of coronary insufficiency and myocardial infarction in the elderly solely from the medical history and clinical findings. One of the most common diagnostic criteria, namely precordial pain, is often absent in myocardial infarction in the elderly. In the Gothenburg study, approximately 60 per cent of those who had previously suffered from myocardial infarction, as judged from the electrocardiogram using the Minnesota criteria, had never experienced any precordial pain and were unaware that they had previously had a myocardial infarction.

It should also be emphasized that the prevalence of chronic bronchitis at advanced age is approximately twice as high in males as in females. At the age of 70 years, the prevalence in Gothenburg was 18 per cent in males and 9 per cent in females, defined according to the WHO diagnostic criteria. This figure must, of course, be related to the prevalence of smoking in the different populations. The prevalence of diabetes mellitus was 5.8 per cent for men and 5.5 per cent for women at the age of 70 years, and 10.6 per cent for men and 5.4 per cent for women at the age of 75 years. In the Finnish investigation, the total prevalence figure was about 6 per cent in those aged 65 years and above. It should be emphasized, however, that present knowledge does not allow a clear demarcation between the increase in blood sugar level known to accompany physiological aging and the disturbance of glucose metabolism occurring in diabetes.

THE PREVALENCE OF DEMENTIA

Organic brain disorders are examples of abnormalities that obviously increase with age. This means, of course, that demographic data within a population will influence the prevalence of these diseases. Since life expectancy in the Scandinavian countries is among the highest in the world, it is of interest to present data on prevalence collected in those countries.

It should be emphasized that the diagnosis of organic dementia is not always accurate; the prevalence figures reported are probably too high. From the experience of the Gothenburg study, it is clear that many patients with pseudo-dementia had been included in the senile dementia group. Moreover, reduction of psychomotor speed as a manifestation of aging is often misinterpreted as being an indication of a decline in other intellectual functions as well. The polypharmaceutical approach to remedying symptoms in the old, which exists in the industrialized countries today, tends to influence intellectual alertness. In addition, there is evidence from several studies that inactivity *per se* can reduce psychomotor speed and creative thinking. The living conditions of some elderly individuals are associated with social isolation and inactivity to a degree that leads to a pseudo-dementia which may be misinterpreted as a symptom of real dementia.

Prevalence figures are available for organic dementia in the Scandinavian countries and the United Kingdom. Roughly, it can be estimated that the prevalence of organic dementia is between 3 and 5 per cent of the population aged 65 years and over.

CANCER AMONG THE ELDERLY

Very few epidemiological studies have investigated cancer incidence among the elderly, but vital statistics from different countries show that age is the single most important factor in the incidence of cancer.

Coping with daily life

In part of the Framingham study,[8] a cohort of elderly people was assessed for their ability to carry out five aspects of daily life: house-keeping, transportation, social interaction, food preparation, and grocery shopping. Only 6 per cent of the cohort (2654 persons aged between 55 and 84 years) were unable to cope in one or more of these areas. The vast majority were self-sufficient, although a quarter were at high risk of developing an inability to cope in one or more of the areas. This large minority are, therefore, potential candidates for future assistance from the social services.

Preventive measures to be developed and/or applied

GENERAL

Hazard control   (largely a function of social policy)

☐ Reduce the amount of the hazard brought into being (e.g. speed limits on roads).
☐ Modify the release of the hazard (e.g. safety controls on energy sources, such as domestic cookers, and other aspects of product design).
☐ Separation (isolation) of the hazard (e.g. pedestrian precincts in shopping areas).
☐ Modify the qualities of the hazard (e.g. maintenance to eliminate uneven floor surfaces).
☐ Increase awareness of the hazard (e.g. illumination).
☐ Hazard resistance (e.g. fireproof fabrics).

PRIMARY

Health promotion (again, largely a function of social policy, with development of appropriate services)

Anticipation of impairment and disability
☐ Screening for sensory impairment.
☐ Assessment of functional capacities in the elderly at home.
☐ Routine pedicure (aid in cutting toenails and avoidance of painful feet).
☐ Adequate diet (availability of small portions for purchase, pricing, meals on wheels, luncheon clubs, etc.); the role of specific dietary components, such as vitamin D, is uncertain because 'normal values' in the elderly are inadequately documented and cannot be extrapolated from younger adults.
☐ Routine dental care (to facilitate assimilation of available diet).
☐ Promotion of communication.

Prevention of handicap and its secondary determination of impairment and disability (strategy of enablement)

☐ Reducing physical contributions to isolation (housing policy and urban design, including sheltered housing).
☐ Recreational opportunities (intellectual stimuli and preservation of physical capacity, including co-ordination).
☐ Social opportunities (clubs, day centres, etc.).
☐ Social attitudes (relevance of elimination of traditional role of elderly as fount of wisdom and experience because of pace of technological change).
☐ Policies related to transport, financial assistance for mobility, energy, architecture, etc.
☐ Policies to help the helpers (financial assistance, services and holiday relief).
☐ Minimizing separation (planned hospital admission procedures to reduce duration of stay and period of dependence, such as by pre-admission assessment clinics).

SECONDARY

Early detection

☐ Monitoring health status to minimize contributory problems (e.g. influences of physical infirmity and brainstem ischaemia on proneness to falls) and to reduce interaction or synergism of health-related adverse experiences on overall functional performance.
☐ Early warning systems (e.g. Caritas, street wardens, role of persons making deliveries, etc.).

Treatment (to stabilize and repair)

☐ Optimum treatment of identifiable health problems (control of problems while avoiding toxic polypharmacy or inappropriate surgery).
☐ Closer liaison to facilitate service co-ordination and mutual continuing education between primary health care staff and specialized care facilities, to promote devolution of care and expertise.

TERTIARY

Rehabilitation

The majority of rehabilitation resources are currently applied to the elderly, but efficiency and effectiveness may not be adequately established. It must be made clear that competence and ability to cope should be the main criteria for assessing an old person's needs (this approach will, however, require new approaches in the education of professionals who deal with the aging).

Exploitation of rehabilitation potential in primary care may not always be adequate (personal contacts with academic departments of general practice suggest that teaching about rehabilitation is too rarely confronted).

Continuing care

☐ Problem of arbitrary administrative separation between health and social services.
☐ Arrangements for terminal care.

Conclusions

Disabilities increase with advancing age, and the elderly are over-represented in the disabled groups. Nevertheless, the data show that only a minority of non-institutionalized elderly people have disabilities that require extensive public assistance and/or support. The elderly are often referred to *in toto* as a 'problem group' instead of a multitude of individuals, many of whom happen to have one or more problems related to their age. Characterizing the 'elderly' as a needy group creates unnecessary negative stereotypes of society's elders. It is fashionable to state that age itself is not a disease, which is another extreme, creating far too positive stereotypes of society's elders. The truth is somewhere in between; some of the elderly create identifiable subgroups which have special public health/medicosocial problems, and these problems should be met, preferably by solutions not leading to institutionalization. These conclusions also indicate the areas of WHO's future concern in health care of the elderly:
☐ to become better acquainted with the problem of links between disability and aging;
☐ to develop and test model solutions for identifiable subgroups of the elderly, who require specific types of support.

References

1. *International classification of impairments, disabilities and handicaps.* World Health Organization, Geneva (1980).
2. Svanborg, A. Seventy-year-old people in Gothenburg. A population study in an industrialized Swedish city. II. General presentation of social and medical conditions. *Acta med. scand.* Suppl. 611 (1977).
3. Isaacs, B. and Neville, Y. *The measurement of need in old people.* Service Studies, No. 33. Scottish Home and Health Department, Edinburgh (1976).
4. Gruer, R. *Needs of the elderly in the Scottish borders.* Scottish Health Service Studies, No. 33. Scottish Home and Health Department, Edinburgh (1975).
5. Kalimo, E. *et al.* Health and health care of elderly populations in Finland. In *Geron XXII yearbook 1978–79*, Societas Gerontologica Fennica, Hensinki, p. 65 (1980).
6. Dahlsted, S. *Slow pedestrians — walking speeds and walking habits of old-aged people.* Report R2. Swedish Council for Building Research, Stockholm (1978).

7. Sourander, L.B. *et al.* A health survey on the aged with a 5-year follow-up. *Acta socio-med. scand.* Suppl. 3 (1970).
8. Branch, L.G. and Jette, A.M. The Framingham disability study. I. Social disability among the aging. *Am. J. Pub. Hlth* 71, 1202 (1981).

5

Delivery and community participation

*Michael H.K. Irwin**

While further scientific advances in the prevention of disabilities are still needed, the immediate problem is the delivery of the existing technology to those at risk. A first priority is that national and international programmes should emphasize the importance and the practicability of preventive measures such as immunization against poliomyelitis, rubella, and measles; improved nutrition; the development of better child-bearing practices and the encouragement of breast-feeding; the prevention of blindness and deafness through the control of infection and malnutrition; and the prevention of home, occupational, and road accidents. They will prove to be more cost-effective than those designed to rehabilitate the already disabled. In giving priority to preventive programmes in the allocation of resources, governments should recognize that national non-governmental agencies and community groups can be mobilized to support these activities as well as to help with rehabilitative services.

In the rehabilitation of the disabled, the principal emphasis should be on the development of self-sufficiency and the active participation in society of the disabled, building on the resources and supportive strength of the family unit and the reinforcement of the value systems of the society, rather than increasing any dependency on specialized institutions which are usually organized along relatively expensive 'western' models and reach very few disabled persons.

At the national level, it is important to recognize that programmes both for prevention and rehabilitation will draw on the resources of, and transcend the responsibilities of, several Ministries. No single Ministry or Agency is concerned totally with prevention or with rehabilitation. During the International Year of Disabled Persons, nearly every country established a special National Committee. In many instances, this was the first time that any co-ordinating structure involving governmental agencies, non-governmental organizations, and even the media, had been established to solve the problems of disabilities; fortunately, a considerable number of these Committees are being reorganized and remaining active beyond 1981. In fact, IYDP is generally regarded, not as an event of one year's duration, but as a beginning of much greater efforts to prevent disabilities and to help those already disabled. An interministerial executive committee is essential to assure the optimum use of always limited resources and to enunciate clearly a

*Senior Adviser on Childhood Disabilities, UNICEF.

strategy and programme of activities most appropriate to the problems of the country concerned. Such an interministerial body should be fully cognisant of the strategies recommended by the United Nations and its Agencies, and by international non-governmental organizations, as well as other external resources, manpower and financial, which potentially could be made available.

The many non-governmental organizations within a country, or area of a country, should relate their initiatives and activities through a single co-ordinating body working directly with the interministerial committee to ensure that the cumulative efforts are additive and not competitive or duplicative. Non-governmental groups often initiate innovative projects which, if successful, can later be expanded and become incorporated into government services. Also, in individual countries, it can be most helpful if an interagency group, consisting of the local representatives of the United Nations organizations, bilateral agencies, and the major international non-governmental organizations, is formed to jointly assist the national co-ordinating committees in the preparation and implementation of suitable projects to improve the existing services for the prevention of disabilities.

Unfortunately, there is little likelihood of a world economic upturn in the next few years. It is very important now to find ways of increasing the ratio between resources and results. As no-one will disagree with the familiar adage that 'prevention is cheaper than cure' or rehabilitation, the emphasis must be repeatedly on better preventive measures. It cannot make economic sense for one-third of children's hospital beds, in developing countries, to be taken by children who have easily-preventable diarrhoeal diseases. Surely it is almost criminal that we allow hundreds of children to go blind every day when vitamin A capsules, costing only a few pence, or even a handful of dark green leafy vegetables, could have preserved their sight?

One vital way to get more benefits for each aid or health pound available is to use an army of paraprofessional workers, backed by more specialized government facilities, to provide the basic services, within the context of a primary health care system, which together will prevent so many disabilities. The use of such community-level workers, who can provide adequate health care for a cost of a few pounds for each person annually, is not only an economic necessity if the needs of the great majority in poor rural communities or urban slums are to be met, but it is also socially more acceptable and appropriate. Ideally, these paraprofessionals come from the areas in which they work. They are trained to serve in their own village or slum, and thus, they answer directly to their constituency. However, because there is now a trend to give more and more responsibilities to community-level workers, due care must be taken not to overburden them with too many activities.

But an even more important factor to multiply the effectiveness of objectives and resources is to ensure people's active participation in the programmes which are designed to help them. In the past, many health schemes failed because, drawn up by experts and officials in the capital to help poor communities, they were never put to the acid test of whether the poor in question actually wanted them. The impoverished these programmes were intended to help were cast in a somewhat missionary image; innocent, simple souls, smilingly grateful for nutrition classes, baby scales, and vaccines. It did not occur to the experts and the officials that those living in poor villages and slums might be suspicious of what was being done for them, and to them, and therefore had little desire to get fully involved. When government agencies, committed to high social aims, sought to deliver programmes meeting 'basic needs' to a passive populace, they ignored the fact that only the active understanding and involvement of each person in his own health care, can bring about the major changes that are necessary to achieve major health goals. Presently, all too often, major aid providers have a strong commitment to centralized bureaucracies — the donor elite tends to work with the recipient elite. Support usually flows to the most articulate, rarely to the poorest, of any society. Political determination is needed to redirect how health services, providing important preventive measures, can reach those in greatest need.

Everybody can be involved; even children. In recent years, in several countries, successful 'child-to-child' projects have been developed in which school-children have been encouraged to concern themselves with the health of their younger brothers and sisters, and have been taught simple preventive and curative activities which are appropriate to each local situation.

Progress towards success in the prevention of many disabilities will require changes in established habits. But people everywhere resist change. For the majority of people in developing countries, who live close to the edge of subsistance and whose only buffer against the ever-present threat of disaster is an unquestioning conformity with a traditional way of doing things, it often seems illogical to make a change for something which an outsider says is 'better'. Their particular way of life has sustained them for generations. Altering a certain lifestyle, as suggested by a foreign expert or an official living comfortably in a capital city, may appear to be too risky. But is this surprising when a particular health programme may appear to be alien to them: when the beneficiaries themselves have not been involved in the planning of such programmes? After all, there are many so-called educated people in Western countries who will ignore 'sound advice' to stop smoking cancer-producing cigarettes or wear safety-belts in cars when driving. Diseases leading to disablement are not going to be eradicated by externally organized mass campaigns but by a growing understanding and involvement on the part of all the groups and individuals at risk.

Community participation can be defined as the active involvement of men and women in the identification of the problems and priorities in a certain area; in the design of strategies to meet those priorities; in the management and evaluation of appropriate activities; and in gaining access to the benefits arising from those activities. This participation is most effective and meaningful where the processes of planning and administration are decentralized to the community level. Most peasant societies have evolved social structures upon which decentralized health services might be built; the challenge is to identify, understand, and work with these existing systems. Experience has shown that the accountability of community leaders to their constituents is much higher than the accountability of centrally placed bureaucrats to the people; provided inputs are discussed in a public forum, it is most unlikely that extensive misuse will occur.

A balance has to be found between 'top-bottom' and 'bottom-up' approaches, between central control on one hand and local initiatives and participation on the other. Such a balance is best achieved through the joint co-operation of the government, non-governmental organizations, and the people themselves. Definitely government function is most effective when it is interrelated with community initiatives in providing needed training, supervision, and the co-ordination of activities and the provision of needed resources not locally available.

Programmes both for the prevention of disabilities and for community-based rehabilitation must emphasize the inexpensive and simple approaches over the complex and institutional. To reach the poor (the majority of people living in to-day's developing countries), any new strategy must both cost less and overcome the barrier of people's exclusion and disinterest. Community participation, in its fullest sense, is the method of allowing people to marshal and channel their energies and abilities to improve their lives. It requires organization and motivation; and the outsider's role is to cajole and inspire, to encourage the community to take over the initiative for its own health programmes. When communities begin to understand what causes disabilities among them, they will be the first to seek to develop activities to bring about prevention. Outside agencies can help the community to establish links with the formal services of government, and offer external financial and technical support if and when the community needs it; but, they should not run, plan, manage, impose, or decide. Through community participation, the poor and the relatively powerless can improve on the condition of their lives, and thus generate improved primary health-care efforts which will prevent more disabilities from occurring.

Communication

*William D. Clark**

Participants at the Leeds Castle Seminar agreed unanimously that the question of communication was crucial for any attempt to prevent disablement. There were two main functions of communication:

1. The creation of the political will to act in the field of disability.
2. The use of modern means of mass communication to reach far beyond the political elite to form, change, and influence the minds, habits and customs of the great masses of population in today's world.

The creation of political will to act for the prevention of disability, it was argued, was a matter of letting people know that there was something you could, comparatively easily and cheaply, do to prevent such crippling conditions as blindness, deafness, or leprosy. The fact that comparatively cheap methods of inoculation against such diseases as measles, diphteria, and poliomyelitis were not known to some health authorities in the poorer countries, even less so to those non-experts who would have to vote funds for any such campaign.

It was agreed that a campaign ought to be mounted to make the possibility of prevention or cure of disabling disease, and the facts about successes in this field, far better known to those who have to make decisions about health and budgetary priorities.

The need to change the habits of masses of people living in traditional ways in order to improve their standard of life, their nutrition, and their health was seen as a key problem in the modern world. For instance, it is established that only a fall in infant mortality will lead eventually to a fall in fertility rates; that the greatest cause of infant mortality is the malnutrition that comes from a mixture of poor diet and infected water. But how to change this in the traditional lives of 2–3 billion peasants living in relatively primitive societies?

The modern mass media (radio, film, and other visual presentations rather than the more elitist methods of the printed word, or the tribal edict) seemed almost designed to fulfil this purpose. There was agreement that the Commonwealth, with its common language and its long traditions of broadcasting exchanges across the frontiers of development, was an ideal unit for experiment in the use of broadcasting as a tool of social progress.

But this is not a simple process of centralized output. From experience it has become clear that it is essential that the social or societal messages must be given by people to whom the recipients can readily relate.

This means, in fact, that more advanced societies can only give advice

*President, International Institute for Environment and Development. Formerly Vice President, International Bank for Reconstruction and Development.

and instruction in these forms of broadcasting; the actual broadcasts need to be done as locally as possible.

But, it was agreed, the passing on of experience and advice on how to achieve agreed health and development goals through better communication was one of the priorities the seminar would endorse.

6

Research priorities

Influenced by the discussion that had taken place on disabilities and their causes, the Leeds Castle participants approached the taxing challenge of identifying priorities for the prevention of disablement with a particular framework in mind. They agreed that it was helpful to distinguish between three major types of situation in which disablement arises. These are:

☐ conditions predominantly associated with the individual's *development*, which include both congenital disorders such as malformation and damage inflicted by birth trauma and the effects of malnutrition;

☐ disorders that follow what can be regarded as an *acute* course, which include communicable diseases and trauma associated with accidents and other injuries;

☐ conditions which are *chronic* in nature and where the etiopathogenesis is generally less well understood; this category includes mental illness, alcohol and drug abuse, much sensory deprivation, and diseases which are often considered to be degenerative such as stroke, ischaemic heart disease, arthritis, and bronchitis.

It was also agreed that the prerequisites for control require that research should be considered in four areas, under the headings scientific understanding, appropriate technology, delivery mechanisms, and secondary and tertiary control. Time did not permit exhaustive review of all these aspects, but the most pressing problems on a global scale were discussed. It is against this background that the following recommendations should be set.

Communicable disease

In considering the causes of disablement world-wide, participants concluded that communicable diseases are the biggest controllable scourge and should therefore excite the greatest concern. The highest priority in research is the development of stable effective vaccines which will provide durable immunity at low cost. This is perhaps especially important in regard to leprosy, and to an improved vaccine against tuberculosis. Where effective immunoprophylactic methods do exist, these are often insufficiently exploited in many developing countries and the costs of such programmes are often the obstacle; operations research on assimilating the means for delivering available vaccines as an integral part of primary health care is therefore also of major importance. Better understanding of vector biology should indicate more effective ways of vector control, and there is also scope for improvement in chemotherapeutic agents. However, at the moment these two latter approaches appear to be more complex and difficult, and probably more expensive as well.

Disorders associated with development

Malnutrition is widespread in many areas, and it is a major underlying cause of disablement. Research is needed into the application of existing knowledge so as to reach those segments of society at greater risk — pregnant women and young children at the most vulnerable period, the time of weaning (between six months and three years). Two areas of particular importance are the development of appropriate combinations of local foods; exploiting simple home or village-based methods of preparation of available nutriments; and means of educating mothers, families, and communities about preferred infant feeding practices, especially during infectious episodes. Also of great relevance is research directed towards fertility control so as to increase interpregnancy intervals (child spacing), which has immediate bearing on the health of the mother and child and hence on the avoidance of disabilities; investment is needed on increasing the application of existing techniques, and on the development of new and safer methods for effective contraception.

Many developmental impairments result from prenatal and perinatal morbidity and appropriate strategies for control need to be explored, particularly in regard to birth trauma and congenital malformations. Work is also needed on the epidemiology of mental retardation in developing countries and, more generally, on the early detection of the condition.

Accidents

The prevention of injury presents a range of options that command further study. These include greater understanding of the epidemiology of home, agricultural, industrial, and traffic accidents; the importance of appropriate technology in the form of product design and testing, especially in regard to safety devices such as implement guards and machinery controls; the influence of human behaviour patterns, including consumption of alcohol and medicaments and the role of regulation by means such as legislation and its enforcement in regard to the application of home safety procedures, protection against occupational hazards, and highway control.

Chronic diseases

Progress in control of the emerging burden of chronic disorders is severely handicapped by limitations in scientific understanding of how these conditions arise. Disorders of the circulatory system, notably cardiovascular disease (including hypertension) and stroke, are an urgent problem, especially as their toll is now beginning to rise in developing countries. There are many fruitful research opportunities to pursue in regard to risk factors, such as the influence of life-style (including diet, and especially salt intake), physical activity, stress, and other elements (e.g. noise).

The Seminar had little opportunity to survey the full range of chronic disorders, but three other aspects were singled out for attention. Chronic lung disease requires further research into the control of industrial and environmental pollution. Better understanding is needed of the processes leading to chronic arthritis, especially of the forms that are commonly regarded as degenerative as there should be considerable scope for prevention. Finally, more work is needed on the influence of agents in the environment; cancer provides an example of where such investment could be very fruitful.

Other aspects of control

Two elements of relevance to economical control of both specific disorders and combinations of disabilities were highlighted at the Seminar. The first concerns recognition of vulnerable individuals, families, and communities so that initiatives to reduce hazard can be instituted. Further research is therefore needed on the means for identifying those at high risk for potentially disabling conditions. The second area complements this by emphasizing the importance of early detection of disabilities and prompt intervention. More research is needed to devise simple detection techniques that can be employed by community workers; to improve effective communication and institution of early treatment at community level; and to develop the role of the family as the basis of the support system.

Another field of concern is the availability of resources not only for the implementation of control measures, but also for the prosecution of research so as to be able to remedy deficiencies. In regard to both of these initiatives the Seminar considered there to be a large and mainly untapped potential of resources and co-operative participation on the part of the community at large, notably through the enterprise of voluntary agencies. Further work is needed to find out how more of this latent capability can be harnessed to the control of disablement.

The remaining topic of major interest is what might be termed the salvage of those already afflicted with severe disablement. It has been estimated that 70 per cent of the world's 'disabled' population is not receiving remedial help, although a United Nations Expert Group has shown that rehabilitation services are in fact an important way of reducing the overall costs of disablement to society. More research is needed to identify which components of rehabilitation programmes have the greatest impact, and how the fruits of such understanding can be made more widely available; the latter is almost certainly another instance of the unrealized possibilities of primary health care. Finally, more study is needed to clarify the role of social policies in various areas other than health and rehabilitation services in promoting the re-integration and assimilation of the 'disabled' into the constructive activity of the community.

7

Prospects for control

*Philip H.N. Wood**

Requirements for control

The development and implementation of control strategies has to rest on disciplined appraisal of the problem it is intended to control. The major components of such an appraisal are:

☐ awareness, so that the problem is perceived (i.e. its existence is recognized);

☐ definition, which includes specification of the nature of the problem and identification of associated attributes that characterize the subpopulation at risk; the latter can provide the basis for screening programmes and preventive intervention;

☐ establishing its priority by examination of the frequency and severity of manifestations of the problem, and of any trends that may be evident;

☐ documenting the consequences to which the problem gives rise;

☐ taking stock of the potential for intervention, which is a function of aetiology and which has to take account of primary, secondary, and tertiary levels of control, and evaluation of the effectiveness and efficiency of such measures.

On the foundation of such an appreciation it should be possible to identify objectives and from these to develop policies, organizational or administrative responses for control. Resource application for these purposes is likely to take two forms. First, measures for intervention will need to be deployed at the most appropriate level, be these concerned with advice or 'treatment'. Secondly, investment may be necessary to make good deficiencies — i.e. research may be called for.

There are three prerequisites for successful control, and acknowledgement of these provides a key for assessing in what areas deficiencies need to be remedied. The fulcrum for control is scientific understanding, even if this is not a *sine qua non*; after all, man learned how to control scurvy long before he knew about vitamin C. Certainly our understanding of many of the major health problems that give rise to disablement is pitifully meagre. For instance, it has been estimated that at present we can account for only 15 per cent of the variance in the occurrence of ischaemic heart disease.

Many hopes for improvement must therefore rest with increasing *scientific understanding*. This is likely to result from complementary progress in two broad areas. The first concerns better characterization

*Arthritis and Rheumatism Council Epidemiology Research Unit, University of Manchester Medical School, Manchester.

of the problems, both as regards definition (nosology) and the circumstances of their occurrence (epidemiology). The second area relates to the processes or mechanisms involved in the production of disease and disablement, and draws on the whole range of unspecialized[1] or so-called fundamental research disciplines. New options for intervention may be expected to emerge from advances in these areas. However, such endeavours may be constrained not only by limitations in the guiding paradigm, but also by moral biases. The Puritan ethic tends to exaggerate individual responsibility, and we are now seeing blame attached to people for some of their health misfortunes. Such a view glosses over the fact that 'this issue is problematical as attempts to promote self-reliance, increased participation in community affairs, and responsibility for individual and family health are based on an assumption that people are in a position freely to choose what course of action they will adopt. However, the options open to an individual may be severely restricted by the powerful forces of socialization'.[2]

The second prerequisite is the development of *appropriate technology* to exploit the understanding. At a very technical level this is illustrated by the way in which total replacement of the hip joint did not 'take off' (i.e. was not carried out on a very wide scale) until Charnley developed both the low-friction arthroplasty and use of a cement for fixation of the implants. But this is too narrow and elitist a view of technology, and the isolation of hazards (including quarantine), means for improving uptake in vaccination programmes, and primary health care all have to be regarded as being in the same category. This global view of appropriate technology makes it clear that investment to make good deficiencies calls not only for applied research on methods of prevention and cure, but also for scrutiny of the organization for delivering the technology by means such as operations and health service research. Selection of realistic objectives is also influenced by technological capacity, as can be illustrated by comparison of the prospects for eradication or control of infectious diseases.[3]

The third prerequisite is *mobilization of the will of the people* to overcome the problem, which can be conceived of on more than one plane. At a societal level this raises political and economic issues, concerning both the application of resources for solution of the problem and a readiness to confront the need for change in institutional arrangements. The latter include not only specifics, such as barriers to access to public buildings, but also broader areas of social policy that determine opportunity; for example, both health and disablement status may be influenced, for better or worse, by the organization and availability of education and employment, and by cultural frameworks that establish the relationship between one segment of society and another.

At a more individual level are not only the initiatives people take in regard to their own health, but also the attitudes and behaviours they manifest towards others. The latter are particularly potent in provoking the more intimate disadvantages associated with disablement, epitomized by the depersonalization resulting from phenomena such as stigma and stereotyping (e.g. the 'Does he take sugar?' syndrome). Attempts to modify these various influences by mobilizing collective will can be described only in very general terms. Debate is obviously called for, so that the relevant issues are examined. This, in turn, will depend on awareness, which indicates that public education has a critical contribution to make — a challenge to develop much broader initiatives in health education than have been evident so far.

Nature of disablement

Disablement is a compound concept, concerned with the consequences of disease and illness. It has been brought to the forefront as a pressing problem by changes in many aspects of health experience.[4] There has been a marked growth in what might be regarded as the hosts, the population vulnerable to disablement. The major force has been demographic change, increased life expectancy being associated with survival of far more people to form the at-risk group. However, the magnitude of the problem has also been added to by technological developments which have facilitated prolongation of life, albeit with reduced competence, for many who would formerly have succumbed from various disorders.

Alteration in the underlying conditions has already been acknowledged, how the solution of one set of problems has served to reveal fresh ones that are more intractable. Thus an alarming burden of chronic disease has emerged as many acute and life-threatening conditions have been conquered. Finally, the milieu in which these experiences occur has also been subject to change. This has been especially noteworthy in regard to responses to the problems. On the one hand increased provisions for health and welfare have been made, and these have been associated with a trend from charitable to insurance or state funding. On the other hand there has been an explosion in expectations, an uneasy amalgam of faith in the potency of science, diminished tolerance for symptoms, demand for egalitarian opportunities, assertion of autonomy, and concern over issues such as death and dying, bioethics, self-help, pain relief, and alternative care.[5]

The time scale of chronic illness is perhaps its most critical property, the feature that compels exploration of the accompanying processes that constitute the consequences. The medical model is concerned with the intrinsic situation, the occurrence of something abnormal within the individual. This is followed by exteriorization of the problem, where someone becomes aware of the abnormal occurrence — be this the affected individual, with so-called clinical disease or with in-explicable symptoms, or a third party, as when a relative perceives a problem or when subclinical disease is detected, perhaps as a result of screening. In turn this experience if objectified as performance or behaviour undergo alteration; everyday activities may become re-stricted, and the whole process can trigger psychological responses, what is referred to as illness behaviour. Finally, these occurrences are socialized as the awareness or altered performance of behaviour lead to the individual being placed at a disadvantage relative to others.

This very brief outline of the development of illness has been con-densed from the *International Classification of Impairments, Dis-abilities, and Handicaps.*[6] Its relevance is that, by identifying ex-periential correlates, it helps establish the foundation for a conceptual model of disablement. The constituents of this model are shown in Fig. 7.1. This framework for appreciation of the consequences of disease is discussed more fully in the ICIDH, although the origins of this development are described elsewhere.[7] In this context it probably suffices to bring out just one other point, that the distinctions made facilitate policy development by clarifying the potential contributions of medical services, rehabilitation facilities, and social welfare respectively.

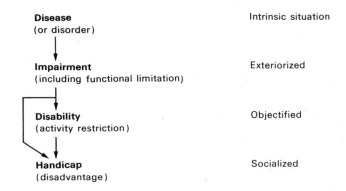

Fig. 7.1.

Control of disablement

A general framework for control strategies has already been described, and this is further exemplified in the appendix. In taking stock of the potential for intervention, it has to be acknowledged that the preven-tion of disablement falls largely within the domains of primary and secondary disease control. It therefore has to be viewed in a condition-specific frame of reference, as has been done in other chapters.

Once a disease or disorder is established, however, and impairments have developed, the challenge then is to try to avert progression down the disablement sequence. In these endeavours a greater commonality is evident, because the general nature of the problems is not so disease-specific. As these aspects receive less attention elsewhere in this book, some further discussion seems to be appropriate.

In regard to disabilities, a hierarchy of intervention goals can be identified. The presence of an impairment indicates that the individual is in hazard of activity restriction. The first level is therefore concerned with true disability prevention, seeking ways of preserving the individual's ability to perform activities unaided and on his own without difficulty. Once difficulty is encountered the goal becomes *enhancement*, by means that may be direct, such as physical re-education or the use of glasses for reading, or indirect, e.g. by increasing illumination or obtaining reading material in a larger print.

If problems are more serious so that activities can be performed only with aid, including that of others, then *supplementation* is the goal — directly, with appliances or a guide dog, or indirectly by environmental adjustment, such as raised marks on control knobs. Finally, if an activity cannot be performed even with aid, *substitution* becomes necessary. This may be accomplished by technology, which can include a hoist, using the radio as a source of news, and listening to talking books, or by adaptation of the environment, of which residential care is the most general example and a blind colony a more specific instance.

It is necessary to consider how impairments and disabilities give rise to handicap. Medical and remedial services tend to concentrate on the individual, but disadvantage arises from interaction with that person's situation. This requires that thought be given to the environment, both physical and social, and to the resources on which the individual can call. For this purpose it is necessary to assimilate both sides of the balance sheet, taking account not only of liabilities, such as the demands that have to be responded to and any limitations on responses, be these physiological or behavioural, but also of assets — these may include the general state of health, underdeveloped talent or potential, and the extent of social support.

The critical property of handicap is its relativity, the discordance between the individual's performance or status and the expectations of the group of which he is a member. The latter element in this disequilibrium, expectations, has not received the attention it merits. I submit that it is necessary and urgent that we should develop public-health strategies directed at modifying expectations and social arrangements in such a way that the needs of a person with a disability would be conceived of differently, and in a less disadvantageous manner. Expectations embody cultural limits for accomplishment or behaviour, which are reinforced by their structural context. These are compounded by preconceptions about how transgressions of these limits should be regarded. Both are amenable to change; the limits may be extended, and tolerance, empathy, and enlightenment can erode prejudice. This is a task that might be included within the ambit of health education. Structural contexts are also constraining, and some general options in regard to these have been indicated in Appendix 7.1.

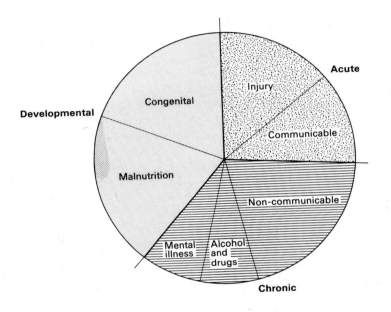

Fig. 7.2. Causes of disability in the world.
(WHO estimates of population affected — 450 million; prevalence one in ten.)

The size of the problem

The concluding part of this appraisal has to take account of the size of the problem. Figure 7.2 shows the world burden of disability, based on WHO's estimates of the proportions accounted for by various of the major causes. I have inflicted my own taxonomy on the data by dividing the whole into three main groups — developmental, acute, and chronic. I find this scheme more helpful when considering the potential for prevention and control. Some 64 per cent of the total is accounted for by developmental and acute problems, a large part of which could be prevented by the application of conventional public-health insights. This conclusion is reinforced when it is recalled that one-third of those affected are children, and that four-fifths of the disabled live in developing countries. Much the biggest challenge, therefore, is to learn more about why what is possible is not put into effect. The remainder of the burden is made up of chronic problems, which are not so

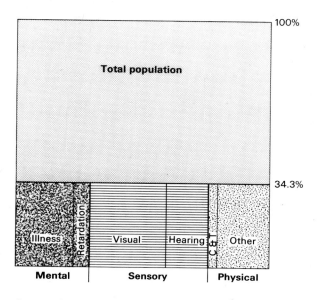

Fig. 7.3. Prevalence of impairment in Great Britain.[9] (C & T = congenital and traumatic.)

straightforward. However, even in this area a great deal can be done in the light of existing knowledge, although research to increase our understanding must also command high priority.

Before proceeding I must give some attention to the quality of the evidence on which these estimates are based. Available data certainly leave a great deal to be desired and the inconsistencies revealed by juxtaposition of estimates from different sources, in an attempt at triangulation, are not encouraging. However, pursuit of precision and accuracy too often becomes an end in itself, especially for the empiricist who is shy of a theoretical context in which to locate his observations. I prefer to follow John Graunt's maxim[8] that very imperfect data, if patiently considered, can teach us something that is good for us to know. I submit that Fig. 7.2 provides sufficient of a diagnosis at the community level to inform the development of global control strategies by WHO, and that it would be almost immoral to be distracted from these priorities by bemoaning the limitations in the data.

Conscious that many of the preventable causes of disablement in the world have been discussed in earlier chapters, it seems appropriate that I should give further consideration to problems nearer home, the spectrum of disability in developed countries. The source data are somewhat better in quality, and so distinctions can be made between different levels of disablement. Figure 7.3 gives an indication of the prevalence

Table 7.1. *Estimated frequency of major classes of impairment and severe disability in adults (aged 16+ years) (United Kingdom)*[9]

Class of impairment	Percentage frequency of occurrence in the general population	
	Impairment (all degrees)	Severe disability
Mental (retardation and illness)	10.1	1.2
Sensory (vision and hearing)	15.9	0.5
Physical (congenital, traumatic, and other)	8.3	1.8
ALL CLASSES	34.3	3.5

of the three major classes of impairment, mental, sensory, and physical, in Great Britain. Any degree of impairment has been taken into account, which is why the alarming figure emerges that 34 per cent of the British population are impaired in some way. The column captioned 'other' represents what is normally referred to as chronic disease. In case it be thought that this snapshot portrays another British malady, it should be noted that a similar order of magnitude obtains in other developed countries.

The key to appreciation of the situation lies in relating these estimates to data on those who are severely disabled and therefore likely to be handicapped in some way. This has been done in Table 7.1 which is based on earlier work of ours[9] and indicates that the risk of severe disability varies considerably between the various types of impairment. For example, whereas one in eight of the mentally impaired are severely affected, the ratio is only one in 31 for those with sensory impairments. Viewed in this light, physical impairments are the most disabling, the ratio being one in five so that these conditions make up slightly more than half of those encountered in the severely disabled.

Taking these various estimates at face value, the evidence suggests that about a third of the population is impaired in some way, a third of those with impairments are disabled to some degree, and a third of the latter experience sufficiently severe restriction in activity as to be handicapped. That offers some feeling for the extent of the problem, but caution is necessary in regard to prevalent misconceptions. Much thought about disablement tends to be dominated by the stereotype of a young adult in a wheelchair, what we have designated as the 'fit' disabled.[10] However, in developed countries the major impact of disablement is evident in the elderly who, because of both their age and the character of the underlying chronic disease processes, have to be acknowledged as frail. Failure to perceive this aspect of reality has allowed medical services, social policies, and research programmes to develop in such a way as to be neither adequate nor appropriate to the nature and scale of the problem.

Trends in disablement are not well documented, apart from what has been said about the emergent dominance of chronic disorders and to note that, in industrialized countries, apart from wartime casualties, trauma is no longer a leading cause (though to acknowledge that is not to gainsay the distressing toll of highway accidents that occur). The importance of age in relation to disablement experience has already been noted. Other attributes of 'the disabled' will vary according to demographic structure and prevalent health problems, and so can scarcely be encompassed in this context; we have considered these aspects for Great Britain elsewhere.[11] The consequences of disablement will similarly vary from country to country, depending on educational and employment opportunities, social welfare provisions, and the like, and so these, too, cannot be enlarged upon.

Conclusions

I hope I have said enough to indicate how ramifications of the problems of disablement have tended to be appreciated insufficiently. With changes in patterns of health experience, chronic disabling disorders are fast coming to dominate the health care scene. It is to be hoped that initiatives stemming from the International Year of Disabled Persons will provide the stimulus not only to confront the challenge more adequately, but also to acknowledge how a large part of the difficulties is intimately related to the very fabric of the societies in which we live. Although the Leeds Castle seminar concentrated on technical opportunities for the control of disablement, and perhaps understandably so, successful implementation of these strategies cannot help but call into question many aspects of the ways in which our lives are structured.

Appendix 7.1.

A model for systematic appraisal of options for control has been referred to, and a potential for confusion over its application to both disease and disablement has been noted. The following table presents selective examples that may clarify usage in the different contexts.

Primary control	(to prevent)
(i) *Health promotion*	(largely a function of social policy with derivative development of appropriate services, notably for maternal and child health)
— disease oriented:	— family planning, antenatal care, breast-feeding, immunization, adequate diet (especially re iodine deficiency and protein/calorie deprivation — legumes, etc.), prophylactic replacement (e.g. salt for workers under thermal stress), and physical activity

— handicap oriented: (enablement)

— urban design (re isolation), educational and employment opportunities, transport policies, social attitudes, etc.

(ii) *Hazard containment* — collective:

(largely a function of social policy)

— repression (prevent creation) : clean water, cassava (cyanide), plutonium production, thalidomide

— reduction (of amount)   : vehicle speed, lead in paint, asbestos mining, atmospheric pollution, noise abatement

— restriction (of release)   : sanitation, pasteurization, onchocerciasis, occupational intoxicants, nuclear waste, product design (fireproof fabrics, automobile styling)

— regulation (of distribution)   : alcohol, tobacco, dietary excess, shut-off valves

— evasion (in time or space)   : quarantine, evacuation, pedestrianization

— separation (material barrier) : product design (implement guards), childproof closures, surgeons' gloves

— modification (of qualities)   : pharmaceutical molecular roulette, product design (e.g. spacing of slats on side of cot)

— awareness (of danger)   : identification, illumination, knowledge (education)

— individual:

— access (e.g. to domestic fires)

— genetic counselling, contact tracing, avoiding ototoxic drugs, prophylaxis (e.g. penicillin for rheumatic fever), identification of risk factors

Secondary control
(i) *Reaction*

(to arrest)
(identifying damage; largely a function of social policy)

— prompt response

— automobile seatbelts, emergency (rescue) services

— screening

— early detection and treatment (e.g. amniocentesis, phenylketonuria, G6PD and thyroid deficiency, hearing impairment, dental caries, glycosuria, blood pressure, glaucoma, anticipatory care of elderly)

(ii) *Stabilization*
— cure:

(countering damage; availability of health care)
— oral rehydration, reattaching severed limbs, otitis media, congenital heart surgery, diabetes,

	pernicious anaemia, endocrine malfunction
— amelioration:	— avoiding impairment and disability by arrest of disease progression

Tertiary control (to repair — control of disablement)

(i) *Restoration* (control of disability — WHO's second level 'disability prevention'; availability of health and remedial care)

— reconstruction: — total hip replacement, post-traumatic cosmetic surgery

— rehabilitation: — remedial services, provision of aids

(ii) *Maintenance* (control of handicap — WHO's third level 'disability prevention'; largely a function of social policy)

— continuing care: — monitoring for deterioration, terminal care

— enablement: — extension of opportunities, vocational resettlement

— support: — welfare provision, assistance, aid to family

References

1. Himsworth, H. *The development and organization of scientific knowledge.* Heinemann, London (1970).
2. World Health Organization Principles and methods of health education. Report on a Working Group, Dresden, 24-28 October 1977. WHO Regional Office for Europe, Copenhagen (1979).
3. Yekutiel, P. Lessons from the big eradication campaigns. *Wld Hlth Forum* **2**, 465-90 (1981).
4. Wood, P.H.N., Bury, M.R., and Badley, E.M. Other waters flow, an examination of the contemporary approach to care for rheumatic patients. In *Topical reviews in rheumatic disorders*, vol. 1 (ed. A.G.S. Hill) John Wright, Bristol (1980).
5. Strauss, A.L. Editorial comment. *Social Sci. Med.* **14D**, 351-3 (1980).
6. World Health Organization International Classification of Impairments, Disabilities, and Handicaps, a manual of classification relating to the consequences of disease. WHO, Geneva (1980).
7. Wood, P.H.N. Appreciating the consequences of disease: the International Classification of Impairments, Disabilities and Handicaps. *WHO Chron.* **34**, 376-80 (1980).
8. Greenwood, M. *Medical statistics from Graunt to Farr.* Cambridge University Press (1948).
9. Wood, P.H.N. and Badley, E.M. Setting disablement in perspective. *Int. rehab. Med.* **1**, 32-7 (1978).
10. Wood, P.H.N. and Badley, E.M. *People with disabilities — toward acquiring information which reflects more sensitively their problems and needs.* Monograph No. 12. World Rehabilitation Fund, New York (1981).
11. Wood, P.H.N. and Badley, E.M. An epidemiological appraisal of disablement. In *Recent advances in community medicine*, Vol. 1. (ed. A.E. Bennett), pp. 149-73. Churchill Livingstone, Edinburgh (1978).

8

Mechanisms for action

The International Year presented to disabled people a unique opportunity to bring their needs and aspirations to public and political attention. The result in many countries was a significant change in attitudes, towards the recognition of the disabled as individuals with the right to participate to the fullest possible extent in social and economic life, to achieve independence through rehabilitation, and, through their organizations, to have a decisive voice in policies affecting their welfare. Imaginative articles in newspapers and magazines highlighted the achievements of the disabled. Radio and television features helped to break down age-old prototypes and described in sensitive detail the problems of disability and the possibilities of physical and social rehabilitation. Films and plays which in previous years would have been seen only by specialized audiences became box-office successes.

Organizations of disabled people made a notable contribution to the success of the Year by understandably grasping the opportunity to achieve an immediate improvement in the lot of the disabled. The International Year itself was conceived within the framework of the United Nations Universal Declaration of Human Rights, and with convincing eloquence and on many political platforms, leaders of the disabled advanced — on behalf of what they claimed to be 'the world's largest majority' — the right to full participation and equality, to rehabilitation and self-determination. In the words of the 'Charter for the 80s' they claimed 'to share the rights of all humanity to grow and learn, to work and create, to love and be loved'.

At the national level, committees were established with the participation of disabled members and with powerful public support, to advocate the needs of the disabled and to formulate national policies. Outstanding support to the International Year was given by the UN and its specialized agencies together with their indispensable partners the international non-governmental organizations. The International Labour Office affirmed the right of the disabled to rehabilitation and employment. UNESCO supported the demand for improved access to information and to integrated education. UNICEF set up its special programme for disabled children in co-operation with Rehabilitation International. Expert committees of the WHO reported generally on rehabilitation and prevention and specifically on the control of many disabling diseases. The Centre for Social Development and Humanitarian Affairs, which was the Agency principally responsible for stimulating and co-ordinating action during the International Year, formulated a world programme of action concerning disabled persons expressed in a 'Global Disability Strategy'.

To organizations of disabled people for whom human rights have often seemed postponable and to the International Secretariat aware of the brevity of an International Year, the initial emphasis on immediate social objectives was entirely understandable. The long-term prevention of disability was from the outset one of the stated objectives and, towards the end of the Year, and in the preparations for the Decade of Action which follows, the balance began to be redressed in favour of prevention. It was in this setting that the Leeds Castle International Seminar — with its emphasis not only on scientific possibilities of control but also on economic advantage and on the need to mobilise political will -- proved to be a timely intervention and a significant stimulus for further action.

In December 1981, at the special plenary session of the UN General Assembly devoted to the IYDP, Sir John Wilson, on behalf of the British Government and as spokesman for the seminar, presented the Leeds Castle Declaration which was subsequently distributed to all delegations. Other speakers on behalf of WHO, UNICEF, and UNDP emphasized the unprecedented opportunity which existed to prevent disability particularly at the level of primary health. A number of national delegations also emphasized that world interest in the disabled should now be expressed in continued action not only to advance the rights of people who are already disabled but also to reduce the burden of avoidable disability, particularly in the developing countries.

This need for long-term preventive action was also powerfully advocated at the Executive Board of the WHO in January 1982. The point was made that if, in the primary health programme, equal emphasis was given to morbidity as to mortality, that programme — at a cost which should not be unacceptable — could offer the best chance of achieving a decisive reduction in the toll of disablement in the developing countries. Three proposals resulted from that meeting: that prevention should be a major objective of the programme of action following the International Year; that an inter-agency consultative group should be established to formulate policies; and that an administrative unit should be established within an appropriate United Nations structure to promote action particularly in the national programmes of developing countries.

At two inter-agency meetings in Vienna, held under the auspices of the UN Centre for Social Development and Humanitarian Affairs, a 'world programme of action concerning disabled persons' was approved for submission to the United Nations General Assembly. In that programme, which initially had been drafted largely in terms of the action needed to eliminate discrimination and to improve facilities for the disabled, the prevention of disability was reinstated as a major objective of the 'Global Disability Strategy' which all member nations are urged to implement as an integral component of their general policy for socio-economic development. The practical measures of prevention to which the Leeds Castle Declaration called attention are incorporated in the world programme of action.

Following the meeting of its Executive Board, the Director General of the World Health Organization agreed to the establishment of an inter-agency consultative group to advise on the practical measures to be taken for the prevention of disablement in existing health and development programmes. This group, which is to be established on the nucleus of WHO's Expert Committee on Disability Prevention and Rehabilitation, will be an effective means of continuing multidisciplinary collaboration and the expansion, to which the Leeds Castle Seminar made a notable contribution, of the conceptual framework of disability prevention.

At the same time, action was being taken by WHO, UNICEF, and UNDP to set up, within the structure of the United Nations Development Programme, an administrative unit to assist with the promotion of action for the prevention of disablement particularly within the national programmes of developing countries. This unit, financed initially from extrabudgetary resources, will complement the work of the Centre in Vienna which, under the 'world programme of action' has general responsibility for continuity of action following the International Year. The location of the new administrative unit within the structure of the United Nations Development Programme, and its close links with three specialized agencies concerned with health and the prevention of disease throughout the world, should enable it to act as an expert resource to government planners and as a constant stimulus to action within those global health and development programmes which can most profoundly influence the prevention of disablement.

The most powerful stimulus to political commitment is likely to come from the communities' own awareness of the scale of the problem of avoidable disability and of the simple measures which that community itself can take to limit the prevalence and consequences of disability. An important step in this direction was taken in July 1982 when an inter-regional meeting in Sri Lanka considered the preliminary results of an initiative by WHO and UNICEF to introduce the concept of 'community based rehabilitation'. The field testing of the manual on 'training the disabled in the community' in ten countries showed that ordinary families, given the right information and support, can both measure the extent of disablement in the community and introduce simple but effective procedures for rehabilitation.

An international effort is now being made to persuade governments to include community based rehabilitation in their development plans. The draft manual, already available in ten languages, is being remodelled and published in the hope that national manuals will be produced in local languages adapted to local social and cultural settings.

This will focus attention on neglected disablement in areas hitherto unreached by rehabilitation services. It will also provide invaluable information of the prevalence and causes of disablement in these areas as the basis for a policy of intervention.

The full results of the field-testing are not yet available, but UNICEF's contention that a high proportion of disablement in developing countries affects children is amply borne out by preliminary figures. These show that in six different countries where the manual was put to work in rural areas, 69 per cent of those detected with disablement were under 16 years of age. As the use of the manual spreads and the results are analysed, much more should become known about the degree of disablement in the developing countries, the categories of disability and the possibilities of preventive action.

Another significant development during 1982 was the preparation of a report commissioned by the Commonwealth Secretariat on policies and programmes relating to disabled people in all Commonwealth countries. This was timed so that the activities of national committees during the International Year could be reviewed. The purpose was to suggest which of the initiatives ought to be sustained by national governments and by Commonwealth co-operation after the Year was over, with special reference to smaller and more remote countries with competing demands on severely limited resources of money and trained manpower.

While much of the report of this survey inevitably reflects the main concerns of national committees with the social and physical rehabilitation of disabled people and particularly with the education and care of handicapped children, it also stresses the opportunities for primary prevention and quotes extensively from the Leeds Castle Declaration. Specific recommendations relate to research into preventable or reversible deafness, to support for the Expanded Programme of Immunization and for the special initiative to extend vaccination against rubella. But the major recommendation recognizes that, since vital decisions about priorities can be taken only at the national level, the follow-up of the International Year will require a continuing mechanism in each country for interministerial action in close co-operation with voluntary agencies and associations of disabled people. Examples are given of statutory councils which can effectively help to devise national policies to make the best use of local resources and of technical co-operation with aid-giving agencies. If such a council is truly interministerial, it can determine the priorities between primary prevention, rehabilitation, education, and employment. If it has a sufficient measure of independence, it can supplement government financing by arousing public support and the participation of the community for specific projects within its programme.

The Commonwealth's facility for communication and for the

exchange of views at all levels of decision-making was mentioned at the Leeds Castle Seminar as a valuable mechanism for disseminating information about practical possibilities for the prevention of disablement. The distribution of this survey report to all Commonwealth governments and the follow-up discussions nationally, regionally, and centrally in the traditional organisms of Commonwealth consultation will help to ensure that the economic as well as the humanitarian sense of seizing opportunities for primary prevention are not overlooked.

In these ways, in the year between the Leeds Castle Seminar and the preparation of this book, there has been useful consolidation of the interest generated by the International Year, clarification of policies and objectives, and the establishment of some mechanisms for action. There has been wider recognition not only that much of the world's disability is preventable or reversible but also that what is required to achieve that result is not a great new vertical programme but rather a shift in the emphasis of global health and development programmes and in the priorities of national programmes. The case which has been made for such action in terms of economic advantage is so convincing that, even in this period of world depression, it has a fair chance of being acceptable and could well improve the acceptability of the larger programmes within which it operates.

The crucial question is whether means can be found of generating and sustaining the political will necessary to convert admirable intentions into real action. At the level of the United Nations, that will has been strongly expressed in the resolutions of general assemblies and executive boards and in the commitment to this cause of the secretariats of the specialized agencies. These secretariats can formulate policies, establish international mechanisms and occasionally (with their indispensable partners the international non-governmental organizations) ventriloquize national calls for action. However, United Nations agencies derive their mandate and resources from national governments: decisive action depends on national commitment and the sense of priority which it expresses.

The eradication of smallpox was achieved only when, following international demonstration of the possibilities, governments adopted the control of this disease as an overriding health priority and set up mechanisms which eventually reached into every community. The West African programme for the control of onchocerciasis, which is now progressing so encouragingly, became possible only when, some 20 years after the means of control had been demonstrated, donor governments provided to a consortium of international agencies the funds necessary to achieve control and West African governments set up an efficient internal administration. The elimination of poliomyelitis is now recognized to be a medical and administrative possibility but, in many developing countries, political commitment to that programme is not yet strong enough to mobilize the fairly modest financial and administrative resources necessary to achieve control. The concept of 'health for all by the year 2000' commands universal support as a desirable objective, but the expanded programme of immunization — which is its main weapon against morbidity — still lacks resources in many countries.

The success of a world plan of action must eventually depend, therefore, not upon a resolution of the General Assembly or the work of the specialized agencies but upon the exercise of political will and the allocation of resources by national governments. The International Year stimulated action to remove psychological and physical barriers to the full participation of disabled people in the social and economic life of their communities. It also called for the provision of better rehabilitation services, not only for the few, as at present, but for more of the majority of disabled people hitherto unreached by them. Most countries have continuing committees or pressure groups to ensure that the impetus of the Year in these respects is not lost and that representations continue to be made to governments by, and on behalf of, adults and children who are already disabled.

The purpose of the Leeds Castle Seminar was to demonstrate the third main aspect of the International Year — the very real possibilities, available now, of relieving the next generation of the present degree of avoidable disability throughout the world. In the words of WHO's expert committee: 'No other single factor can contribute as much to diminishing the impact of disability as first-level prevention'. The message of the seminar, elaborated in this book, is that disablement with all its consequences of wasted resources and frustrated lives need not necessarily be an inescapable part of the human predicament. Prevention on an unprecedented scale and at no great cost is one of the options available to the international community during the next twenty years.

Index